HISTORY, PHILOSOPHY AND SOCIOLOGY OF SCIENCE

SOCIOLOGY OF SCIENCE

Classics, Staples and Precursors

HISTORY, PHILOSOPHY AND SOCIOLOGY OF SCIENCE

Classics, Staples and Precursors

Selected By

YEHUDA ELKANA
ROBERT K. MERTON
ARNOLD THACKRAY
HARRIET ZUCKERMAN

Universities and Scientific Life in the United States

BY

MAURICE CAULLERY

ARNO PRESS

A New York Times Company

New York – 1975

Reprint Edition 1975 by Arno Press Inc.

Reprinted from a copy in
  The University of Illinois Library

HISTORY, PHILOSOPHY AND SOCIOLOGY OF SCIENCE:
Classics, Staples and Precursors
ISBN for complete set: 0-405-06575-2
See last pages of this volume for titles.

Manufactured in the United States of America

———◆———

Library of Congress Cataloging in Publication Data

Caullery, Maurice Jules Gaston Corneille, 1868-1958.
    Universities and scientific life in the United States.

    (History, philosophy and sociology of science)
    Translation of Les universitiés et la vie scientifique
aux États-Unis.
    Reprint of the ed. published by Harvard University
Press, Cambridge, Mass.
    1. Universities and colleges--United States.
2. United States--Learned institutions and societies.
I. Title. II. Series.
LA226.C42 1975   378.1'00973   74-26257
ISBN 0-405-06585-X

Universities and Scientific Life in the United States

BY

MAURICE CAULLERY

PROFESSOR AT THE SORBONNE
FRENCH EXCHANGE PROFESSOR AT HARVARD UNIVERSITY, 1916

TRANSLATED BY

JAMES HAUGHTON WOODS

AND

EMMET RUSSELL

"The world has been remade in the last half-century."
CHARLES W. ELIOT

CAMBRIDGE
HARVARD UNIVERSITY PRESS
LONDON: HUMPHREY MILFORD
OXFORD UNIVERSITY PRESS
1922

TO MY FRIENDS AT HARVARD

AND IN PARTICULAR

TO

GEORGE HOWARD PARKER

PREFACE

THIS book is based on observations and impressions which I gathered during a stay of five months in the United States.

As a biologist, I describe the university landscape above all from the scientific and more specially from the biological point of view, but with the design of making the whole of it understood and of setting it into the general framework of contemporary American society.

During the second half-year of 1915–16, I had the honor of being Exchange Professor at Harvard University. And my first word here must be to affirm, once more, the very great utility of exchanges of professors between French and American universities. They are among the most efficacious means of helping the two countries to know, esteem and aid one another. The great mass cannot cross the Atlantic; but if the educators of youth have done so, they may help to dissipate many prejudices. They are almost bound, it seems to me, not to keep to themselves the experience acquired, however incomplete their observations may often be. That is what has determined me to write the following pages. I wish that they may make better known in France an aspect of American democracy, which is not that under which we are most commonly led to look at it, and also that they may emphasize the efforts which the immediate consequences of the war imperiously oblige us to make without delay.

It is my duty — and a very pleasant one — to inscribe, at the beginning of this book, my best gratitude for the welcome I received in America. American hospitality was shown me from the time I set foot on New York soil. On my arrival in Cambridge, President Lowell received me in his house, and my first impression of Harvard was that of the simple cordiality which is the charm of the Harvard community, and which unites all its members, from the president down to young freshmen. Everywhere, in New York, Boston, Baltimore, Princeton, Yale, Chicago, and at San Diego on the Pacific, I found friends and colleagues to welcome me with the same affectionate eagerness.

Likewise I received favors on the part of learned bodies. I felt particularly the honor which the American Philosophical Society and the National Academy of Sciences did me in inviting me as a guest at Easter 1916 to their meetings in Philadelphia and in Washington.

I have also to thank the clubs — particularly the Colonial Club at Cambridge and the Harvard Clubs of Boston and New York — which, by opening their doors to me during my entire stay, added to its ease and increased its delight.

My colleagues at Harvard, especially those of the department of Zoölogy, welcomed me with an eagerness which the tales of my predecessors had made me expect, but which touched me none the less. At Harvard, they know how to make the newcomer forget, from the first day, that he is a stranger, and to give him the illusion of being a regular and permanent member of the university. Friends watch attentively to foresee the least wishes of the guest, and to remove every difficulty.

And they exercise their ingenuity in making his stay constantly agreeable. I dedicate this book to the delightful memories of these firm friendships.

I understood, through my own experience, what my colleague and friend Paul Marchal wrote recently, in regard to a scientific journey to the United States in 1913, and in particular regarding a stay at Cornell University. "One must have lived for several days," he says, "in the atmosphere of this ideal society of the arts and sciences, in order fully to enjoy its charm, and to understand its harmony, which call to mind the picture of the Future City of Henrik Anderson. One realizes then to what a profound error European travelers are the victims, who estimate American life and civilization, by judgments formed upon the overwhelming impressions which they have felt in the whirl of the great business thoroughfares of New York, or from visiting the famous Stockyards section of Chicago." [1] It is in fact a profound impression of idealism that one brings back from American university circles.

In 1916, during the months when the battle of Verdun was going on, the meaning of it to a Frenchman was singularly reënforced by the warm sympathy which he felt in the unanimity of the American intellectual class for the cause of France and the heroism of her soldiers. He felt himself in the midst of friends more than one of whom regretted not yet being an ally. And he carried away the precious conviction that sincere American

[1] P. Marchal, *Les Sciences Biologiques Appliquées à l'agriculture et la lutte contre les ennemis des plantes aux États-Unis.* Paris (Lhomme), 1916, p. 252.

feeling and the American heart were won for his country, that the best people in America justly appreciated the extent. the purity, and the nobility of the sacrifice stoically undergone by the youth of France, for the salvation of civilization and liberty.

MAURICE CAULLERY.

PARIS, June 1917.

PREFACE BY THE AUTHOR TO THE TRANSLATION

AT the beginning of this English edition, I wish to
express my hearty thanks to my translators, and
especially to my good friend, J. H. Woods. I am glad
that in this form these impressions of my journey will
reach wider circles of American life.

But I should like to warn the reader against errone-
ous conclusions that might be drawn from the book. It
was written for the French public. One should not be
surprised to find many details which seem superfluous
to Americans. And on the other hand, when I spoke of
France and made comparisons I presupposed among my
readers a general knowledge of academic and scientific
life in France and confined myself to allusions. Because I wished to stimulate public opinion, I insisted
almost exclusively on points or reforms that seemed to
me desirable. The result is that only the defects of the
French institutions seem to be noted — which exist in
all countries not excepting the United States — while
this impression does not appear to be counterbalanced
by the solid qualities which our higher education does
actually possess. I should be distressed if the reader,
heedless of the point of view from which the book was
written, would regard it as a general criticism of French
methods. Our traditions often impose upon us heavy
chains. But they have also fertile educative qualities.
Those who have a true knowledge of France, who judge
her without prejudice, can appreciate the clarity, the

solidity, and the refinement in her; and also the high-mindedness which is at the foundation of French mentality. To this Professor Barrett Wendell has himself borne witness. As to the spirit of discovery, it has shown itself, time and again, under conditions the more significant, because the material aids at the disposal of investigators left much to be desired.

M. CAULLERY.

Paris, December 1920.

NOTE BY THE TRANSLATORS

THIS translation was begun in Paris during the war. During the disorganization of the system of transportation the manuscript in its final form disappeared somewhere between Havre and New York. The second version is accordingly much delayed. But it seems unwise to add to the original any attempt to describe the violent oscillations to which American universities have been subjected since 1916. The book thus remains unchanged, a picture of academic life in the United States before the great upheavals of the war, and a pledge of comradeship between French and American universities in the years to come. Thanks are due to Dr. Raphael Demos for assistance on the last pages.

JAMES HAUGHTON WOODS.
EMMET RUSSELL.

MAY 10, 1920.

CONTENTS

PART I

THE UNIVERSITIES

CHAPTER I

Colleges and universities. Recent development. The principal
universities. Private and state universities. Denominational and
undenominational universities.

CHAPTER II

The classical college and the Bachelor's degree. Its evolution in the
nineteenth century. The elective system. The professional schools.
The introduction of scientific research and the graduate schools.
German influence. The equilibrium between the college and the
superadded parts.

CHAPTER III

The campus. Harvard: the Yard and the various additions. Columbia. Princeton. Berkeley. Cornell. Contrast with French
universities.

CHAPTER IV

Harvard, the Corporation and the Board of Overseers. Part played by the Alumni. Other universities. Trustees and Regents. The President. His powers and position.

CHAPTER V

General conditions of the career. Moral and material *desiderata*. Excessive burden of teaching. Insufficient participation in the management. Precarious guaranties. Advances in the career. Salary. Retirement pensions. The Carnegie Foundation.

CHAPTER VI

The classical college (undergraduate). Admission. Organization of studies. Departments. Coördination of courses. Examinations and graduation. College life. Social and collective life. The dormitories. Clubs and fraternities. Sports and athletics. Various associations, dramatic societies. The general results of college studies.

CHAPTER VII

Prevalence of coeducation in the western universities. Its still exceptional character in the eastern. Women's colleges. Parallelism of studies. Social results. Education and the race problem.

CHAPTER VIII

Relations with the college. Development. Degrees. Master of Arts. Doctor of Philosophy. The doctorate in the principal universities.

CHAPTER IX

CHAPTER X

CHAPTER XI

CHAPTER XII

PART II

THE SCIENTIFIC RESEARCH

CHAPTER XIII

CHAPTER XIV

CHAPTER XV

CHAPTER XVI

CHAPTER XVII

CONTENTS

CHAPTER XVIII

CHAPTER XIX

UNIVERSITIES AND SCIENTIFIC LIFE
IN THE UNITED STATES

∴

PART I

THE UNIVERSITIES

CHAPTER I

THE PRINCIPAL UNIVERSITIES

Colleges and universities. Recent development. The principal universities. Private and state universities. Denominational and undenominational universities.

IN order to form an idea of scientific life in the United States, one should study first the universities. Although the scientific effort does not all come from them and although there is even a tendency to organize outside of them the most powerful institutions and those especially destined to promote discoveries, still they remain at the present time the great centres of research, and they are the environment in which future workers are trained. The productive capacity of the country rests therefore upon them; on their good qualities, on their defects, depend the fecundity or the deficiencies of American science. There is, then, an evident interest in making a study of them first, and in revealing their spirit.

And also, they are so different from our own, so linked, as is natural, with the whole of American society, and with the historical conditions of its development, that a complete description is necessary in order to understand them and to analyze their part in the contemporary scientific movement.

Like everything in the United States, they have passed, in the half-century since the War of Secession, through a phase of marvelous prosperity and development. Especially in the last thirty years, this movement has been accentuated. The proof of it will be found in the figures which I shall have occasion to cite in the course of the following chapters. This development has been extraordinarily rapid and consequently hasty. It has taken place in perfect freedom, in an independent manner in the different parts of the country, and not with the uniformity that a central power impresses on the institutions of countries like ours. One feels very strongly that all this has by no means arrived at equilibrium, no more than the cities themselves.

The American university is very broadly conceived. In 1865 Ezra Cornell founded at Ithaca, N. Y., the university which bears his name, and which has become one of the most important in the Union. "My intention," he said, in a phrase which is now the motto inscribed on the seal of this university, "is to found an institution where any man may be instructed in any subject." That is a program as immense as it is generous. It could be only partially realized, but it expresses the present idea of the university. It was, moreover, in the main that of the French Encyclopedists of the eighteenth century, which our Revolution dreamed of realizing without being able to do so.

In principle, the American university considers that nothing is foreign to it, and it offers a diversity of teaching and of schools infinitely greater than the traditional five Faculties (theology, law, medicine, sciences, and letters) of the universities of continental Europe.

In fact, it is the juxtaposition of three principal elements, of which one, the classical college, is historically fundamental. On this college there have come to be superimposed, on the one hand, a higher school of disinterested studies and of scientific research, the Graduate School of Arts and Sciences; on the other hand, the so-called professional schools, furnishing the necessary knowledge for the more or less learned careers, law, medicine, the evangelical profession, and likewise for all the industrial, commercial or agricultural callings. In short, the university aspires to train the leaders in all branches of social activity. We see, then, that it has, through this program, a very vast contact with the whole of the national life.

The college remains the framework of the university. It partakes of the character of our secondary education almost as much as of that of our higher instruction; it is a hybrid between them. Socially, it is the chief element. Its spirit, consequently, is something that one must know. Finally, it constitutes in many cases, by itself alone, the whole institution. There are now, in fact, in the United States, nearly 600 universities or colleges,[1] forming, from the point of view of their material importance, of the elevation and diversity of their studies, a very continuous scale. Quite a number of them are, in reality, institutions for no more than secondary education. All tend to enlarge and to resemble true universities as much as possible. There is a warm competition among all; thus they reflect that spirit of "bigness" which impregnates all American life.

These 600 universities and colleges represent a considerable student population. It numbers at present

---

[1] The report of the Commissioner of Education gives the statistics of 596 colleges and universities in 1912–13, and of 567 in 1913–14. Of this latter number 93 are state or municipal establishments, and 474 are private institutions.

between 200,000 and 300,000, and the figures below show with what rapidity it has increased in less than thirty years.

Years	Men	Women	Total
1889–90..........	44,926	26,874	65,800
1900–01..........	75,472	38,900	114,372
1913–14..........	139,373	77,120	216,493

In spite of their extreme inequality, there is nevertheless a rather uniform general spirit among them, which impregnates the youth who frequent them, and which in a society so heterogeneous as the present United States, is an important factor of unification.

Of course it would not be possible to study here all the American universities, and that would have no interest. The largest and the most perfect alone are important; for the others try to follow their path, and it is in the first alone that one can speak of a true scientific life.

As there is no administrative bond between all these institutions; and as they live and develop in an entirely independent manner one from another, they offer at first sight a great diversity. In reality the resemblances are much stronger than the differences. Therein is a striking example of the influence of the environment and of what biologists call convergence. Common surrounding conditions have resulted in making them uniform, in a large measure.[1]

[1] Nevertheless you must not think there is an identity among them. One can get a good idea of their individuality, and at the same time of their general traits, from the very interesting book by E. E. Slosson, *Great American Universities*, New York, McMillan, 1910.

First, we must distinguish two great groups, independent universities and colleges, and state universities.

The independent universities, which are frequently designated under the name endowed universities, are private institutions, administered entirely by themselves, after the fashion of an industrial or commercial society, by means of a council, generally called a board of trustees. Their resources come from tuition paid by their students, from donations, and from the income of their funds previously consolidated, or endowment.

These private institutions are situated, for the most part, in the eastern United States, that is to say, in the old part, in the states which constituted the thirteen English colonies of the eighteenth century, and which today represent the traditional part of the country, that which is the depositary of English civilization, and which, until now, has given its impress to the rest of the nation.

The most important are the following: in the first place, the oldest, Harvard, at Cambridge, one of the cities which surround Boston and form now a unity with it. An uninterrupted tradition links the college, founded in 1636, with the present university. It is Harvard, besides, which has created the whole tradition of the American college, and after which the younger colleges are modeled. Up to the present it has, almost always, been at the head of the intellectual movement in America, showing the way in most of the transformations which teaching has undergone. This rôle has been assured to it, during the last half-century, in large part by the foresight and boldness of the president who directed it from 1869 to 1909, Mr. Charles W. Eliot, the greatest American authority on matters of education.   Harvard has at present about 5000 students.[1]

Yale, at New Haven, Conn., the rival of Harvard in American university traditions, dates from 1701, and has also had a large part in the scientific progress of the country. It has had among its professors the geologist Dana, the palaeontologist Marsh, the physicist Gibbs. The American Journal of Science was founded at Yale in 1818. Today Yale numbers more than 3000 students.

Three other of the most important universities were founded in the eighteenth century; that of Pennsylvania, in 1740, at Philadelphia; Princeton, in 1751; and Columbia (under the name of King's College) at New York, in 1754.

The University of Pennsylvania has grown parallel with the city of Philadelphia and was one of the first to diversify itself by the addition of special schools. It has today over 5000 students.

Princeton University, situated in the city of that name, in the state of New Jersey, is that one of the large universities which has departed least from the older form of the college; hence its relatively small number of students, about 1500.

Columbia has recently taken, with New York, a prodigiously rapid spurt, materially and scientifically. It remained a relatively unimportant college until about thirty years ago, when it took the name of university, in 1891. Since then it has merged with several special institutions of New York, has become diversified in the extreme, and is today, with its 6000 students,

[1] This figure, like the following ones, does not include the summer schools, as in the table on page 129.

its enormous resources,[1] the strength of its professorial staff, and the level of its studies, one of the most powerful universities of the world.

Several other large private universities, in contrast with the preceding, are of recent creation.

Johns Hopkins, founded in 1875, at Baltimore, thanks to a legacy of the benefactor whose name it bears — a bequest whose amount, $3,500,000, seemed enormous at the time — has played a chief rôle in American higher education, although materially it is rather small. It was planned in a radically different fashion from the ordinary college, as an establishment for true higher learning, for its rôle was to encourage and organize original scientific research. And it has filled this part in a brilliant manner, and has also contributed above all to the elevation of medical instruction. But from having withdrawn itself from the usual conditions, Johns Hopkins has been deprived of the great afflux of receipts and gifts which goes to the other universities, and it has fallen now into rather serious financial difficulties. One may see, moreover, that in the table on p. 147, the figures relating to it are small. Today it has scarcely a thousand students. Its capital is $6,265,000.

---

[1] These are the consolidated interest bearing funds of the different universities:

Harvard	$22,000,000
Yale	15,380,000
Pennsylvania	5,000,000
Princeton	5,000,000
Columbia	33,000,000

These figures, taken from the report of the Commissioner of Education for 1913–14, do not include the value of lands, buildings or equipment, but only the liquidated funds or endowment.

Cornell University, at Ithaca, N. Y., dates from 1865. Founded by Ezra Cornell, and admirably situated on a wooded plateau where it covers no less than twelve hundred acres, it has become one of the most interesting universities, in particular from the point of view of biological sciences, and of their application to agriculture. It numbers more than 5000 students, and its consolidated funds are more than $14,000,000.

The University of Chicago, and Leland Stanford Jr. University at Palo Alto, California, represent the type of independent universities in the West. The University of Chicago is today of the very first rank, in numbers (more than 6,000), in its ample equipment and laboratories, in the high character of its advanced instruction, in the composition of its faculty, and in resources (its productive capital is over $18,000,000). It has been built chiefly from gifts of Mr. J. D. Rockefeller, which have amounted to $25,000,000.

Leland Stanford, which bears the name of its founder, from whom it has received $30,000,000, has been equipped in magnificent fashion. It suffered much from the great earthquake of 1906, which partly destroyed it.

It would be proper to mention still other private universities, besides the preceding. I will limit myself to naming one, very small in number of students, intended to be an institution where research in pure science was to be done by men who should be regarded as "fellows" rather than as students. It is Clark University, founded in 1887, at Worcester, Mass. It has, like Johns Hopkins, passed through difficulties which have not yet ended.

The state universities are differing from the preceding in origin, and in many respects, in spirit. They draw their resources, not from individuals, but from the state. Each western state has, in a general way, its university, which it supports in a very liberal fashion, through its general budget. More than eleven of these universities have subsidies of more, and often much more, than one million dollars.

The origin of most of them goes back to the Morrill Act, passed by Congress in 1862, which gave to the several states considerable areas of public lands, the income, or the proceeds from the sale of which must be devoted to education, principally to the teaching of agriculture and the mechanic arts.

Thus arose the Agricultural and Mechanical Colleges, most of which, by expanding, have become the present state universities. Some have absorbed an already existing college; that was the case in California; others have been incorporated into a university properly so-called. That is the case of Cornell University, in the state of New York, which has, consequently, an intermediate character between private and state universities. Some, in the East, have remained independent, under their original name, like the Massachusetts Agricultural College, at Amherst, which has remained specially biological and agricultural.

Through their origin, the state universities have had at the beginning some very utilitarian tendencies. They have, before all else, striven for practical application and teaching. Real culture has only little by little made a place for itself in them, and is still often rather cramped, and much of the teaching smacks of the soil.

Being sustained by the state, they are more democratic in spirit, and more open to all classes, by the mere fact that their teaching is free, at least for citizens of the state in which each of them is established. Their student population is large, chiefly because they are less exacting as to the knowledge required of their pupils at entrance. But as they grow larger, they tend to approach the private universities of the East in classic tradition, and at the same to rise toward pure scientific research.

Here are a few summary remarks about the most important: The oldest, that of Virginia, founded in 1819 by Jefferson, with views which were far in advance of the time, has been retarded in its development, like all the South.

The University of Michigan, at Ann Arbor, dates from 1841. It is one of those that have attained the highest level and the most considerable development. It has nearly 6400 students.

The University of Wisconsin, at Madison, was founded in 1849. It has today 5000 students. It is in a period of rapid development and has shown a remarkable audacity in the breadth of its program, especially in respect of popular instruction.

The University of California, magnificently situated at Berkeley, on the slopes which border, on the east, San Francisco Bay, facing the Golden Gate, has become one of the largest in numbers (more than 6000 students), and one of the most interesting through its teaching and its publications.

The University of Illinois, at Urbana, has an almost equal importance (5000 to 6000 students — it had but 500 in 1890), and has laboratories very broadly planned.

The University of Minnesota, at Minneapolis, has about 4500 students.[1]

These are the principal state universities, and one may see how ample their resources are. I should mention, besides them, those of Missouri, Iowa, Ohio, and others.

There is without question a group rivalry between private and state universities. Under the similar conditions of the environment, they are coming to a very general resemblance, but their tendencies are nevertheless divergent. The private universities of the East have hitherto represented real culture in an incontestable fashion, and have shown the way. The state universities, by their origin and tendencies, have accelerated the incorporation beside classical subjects, and the development in advanced instruction, of the applied sciences much needed in modern society.

The distinction between state and private universities is the most important, and is the one which I wish to point out here. I shall note, however, in a very brief way, that among the private establishments, some are like the state universities, without allegiance to a particular sect — undenominational, as they are called — while the others are under the control of the church which has founded them. All the large universities belong to the first category. In the second, a special group is constituted by Catholic establishments, and in

[1] Here is the summarized table of the subsidies which these universities have received from their respective states, in 1913–14.

	For New Equipment	For Current Expense
California	$350,000	$1,220,000
Illinois	650,000	1,686,000
Michigan	350,000	1,088,000
Minnesota	941,000	1,420,000
Wisconsin	343,000	1,811,000

particular the colleges of the Jesuit order. These institutions have a history of their own, quite independent of the evolution of American universities properly so-called, and I shall leave them completely aside.

The sectarian origin of many colleges, and the caprice of private foundations, explain how the same city may have several universities, a fact which does not fail to surprise us at first. To cite but a few examples, Washington has no less than four — George Washington University, the Catholic University of America, Georgetown University, likewise Catholic, and Howard University for negroes; and three more colleges. New York, besides Columbia, possesses another large and important institution, New York University, large municipal colleges (City College for men and Hunter College for Women), a Catholic university (Fordham), without counting the colleges of Brooklyn and the school of medicine of Cornell University. Philadelphia has likewise, besides the University of Pennsylvania, several colleges, one being Catholic and another Jewish. Chicago, besides the university of that name, is the seat of the Faculties of medicine of Illinois and Northwestern Universities, of an important technological school (Armour Institute), and finally of two Catholic universities. Boston, besides Harvard across the river, has Boston University, which is sectarian, and the Massachusetts Institute of Technology, the greatest engineering school of the United States. Tufts College and Wellesley College are in its immediate neighborhood.

From the point of view which occupies us here, and outside of two or three particular cases, the establishments attached to a particular sect have no real importance, and it will suffice us to consider hereafter the universities on which I have commented in this first chapter.[1] We are now going to study their life under its different aspects.

[1] I shall only be concerned with the universities of the United States. Nevertheless I shall indicate that the Canadian universities are developing in a quite parallel fashion, and have, moreover, very close relations with those of the Union. The chief are McGill University at Montreal and the University of Toronto.

CHAPTER II

THE BEGINNINGS: FROM COLLEGE TO UNIVERSITY

The classical college and the Bachelor's degree. Its evolution in the nineteenth century. The elective system. The professional schools. The introduction of scientific research and the graduate schools. German influence. The equilibrium between the college and the superadded parts.

ALTHOUGH America is the New World, the universities as we see them today, are the resultants of very ancient traditions and customs.

All that has been introduced recently has been set into the original framework, and adapted to the past. In France it would not have been done so. In a strongly centralized and bureaucratic country like ours, university institutions were created from almost nothing at all by Napoleon. But this method of his is not the most favorable for imparting to the universities real vitality. Yet a progressive evolution is a fatal necessity, in case of private institutions, like the American universities of the East, the first in time and the models for the others. They reflect a society and its history, and are a heritage from English life.

Thus in order to understand them, we must first recall their origin. As has already been said, the university is a metamorphosis of the college, or rather an epigenesis of it. The college still subsists, it is the axis about which the other parts have been articulated; equilibrium is not yet completely established between them and it. It is therefore the story of the college which must first be given.

It goes back to the first years of the New England colonies. In 1620 the *Mayflower* brought the Pilgrims to the Massachusetts coast. Sixteen years later, in the neighborhood of the city which is now Boston, within the town which was beginning to rise on the left bank of the Charles River, they created a college, like those of the mother country. And as many of the colonists came from Cambridge, they gave the same name to the new town where they placed their establishment. Therefore it is indeed the English college of Cambridge and Oxford which is the prototype of the American College. Harvard took the name of the first of its benefactors, the Reverend John Harvard, who died in 1637, leaving his library and a sum of £600, the first contribution to its funds.

It was only in 1701 that the second college of the American colonies was founded, Yale, in Connecticut, at New Haven. Princeton, Columbia, and Pennsylvania date from the middle of the eighteenth century. There were eleven colleges at the time of the War of Independence, eleven others were founded between that event and 1800, 33 from 1800 to 1830, 180 from 1830 to 1865, and 236 from 1865 to 1900.

These eastern colleges, today of an entirely private character, were in origin products of the community. Harvard was created by the Massachusetts General Court; it was governed by a committee comprising at first the governor and lieutenant governor of the commonwealth, and clergymen of the towns near Boston. This committee has been subdivided into two since 1650, one composed of seven persons, including the president and the treasurer of the college, which has been perpetuated down to our time, under the name of the Corporation (or the President and Fellows of Harvard College), and it has retained all the powers of initiative, of executive, and of finance. The other has become an advisory council, which is today the Board of Overseers. This duality at Harvard is moreover a special exception. The true and more general evolution has consisted in the gradual elimination from these councils of the representatives of the government who figured in them *ex officio*.

From the beginning, following the English custom, the colleges could acquire property, and they became more and more independent in the management of it. At their distant origins, the first universities, though private today, were thus in a certain measure state institutions.

The essential function of these colleges was, and remained until after the opening of the nineteenth century, the intellectual training of clergymen. The members of their councils were for a long time almost exclusively official personages and ministers of religion. The great majority of their students went into the Church; 75 per cent at Yale for example (the proportion today is 3 to 4 per cent). These colleges long remained of very modest dimensions. About 1830 Harvard numbered 10 professors and about 200 students. Columbia had 6 professors and 125 students. That also explains why the majority of these institutions, chiefly theological in design, were founded by or originated from churches. The clergymen were moreover, with the lawyers and to some extent the physicians, the only classes in American society of that time who possessed anything like a classical education.

College teaching was conceived with a view to these needs. The pupils lived there together, as at Oxford and Cambridge. Their studies were all of the same character and for the most part classical. They dealt with English and the ancient languages (Greek, Latin, Hebrew) with a little mathematics. They were little by little stereotyped into an unchangeable program which was called the curriculum. They were spread over four years, designated by the traditional names of freshman, sophomore, junior, and senior. At the end of four years the student left college with the diploma or degree of Bachelor of Arts, A.B. After another year's study they obtained the A.M. That was the *summum* of liberal education in America until a half-century ago. The subjects of the curriculum had acquired a sort of nobility in contrast with all others.

Only in the nineteenth century did the development of industry lead gradually to the creation of special schools, preparatory to the professions. Thus were organized one by one, at Harvard, at Philadelphia, and at New York, schools of medicine, of law, and of theology; but they remained a long time rudimentary.

Schools of applied sciences were also created. About the middle of the nineteenth century, Harvard and Yale organized, to this end, in close connection with the college, yet distinct from it, the former the Lawrence Scientific School, the latter the Sheffield Scientific School, where studies lead to the degree of Bachelor of Science, S.B. For a long time — even today — this degree had not the prestige of the A.B.

As has already been said, it was because the colleges refused to give to applied scientific studies a place which had become necessary, that Congress determined to found the colleges of agriculture and mechanic arts.[1] It was likewise in response to this same need that a series of engineering schools was created, independent of the colleges; and in particular, in 1865, at Boston, the Massachusetts Institute of Technology, which grew rapidly.

However, as scientific instruction was making its way in the college itself, the old curriculum broke down, studies were diversified, and the elective system was substituted for the preceding uniformity. Mr. Charles W. Eliot at Harvard, contributed much to this new development. The colleges began to organize very varied courses of instruction. Each student chooses from them, according to his tastes and his needs, a given number of courses, extending over four years, and the whole of which formed the minimum required for the Bachelor's degree. This system has certainly been carried to excess. The colleges, in friendly emulation, compete in offering programs as broad and varied as possible, but without much coördination, and the students' choice was often made rather with a view to the least effort than to the coherence and strength of their course.

Today, at least in the good universities, there is a check on this freedom of choice. It is regulated. There are obligatory fundamental studies, especially in the first years of college. But the possibilities of choice remain very vast; the more so because, even in secondary education, at the high school, they are rather numerous.

Parallel to the diversification of college studies, schools furnishing the knowledge necessary for definite

[1] See ch. x, p. 116.

professions grew up beside the college. The teaching in them is no longer entirely disinterested and purely cultural, as in the classical college. These schools are generally called professional schools or colleges. They are schools of theology, law, medicine, colleges of engineering, agriculture, or commerce. The unity of the college has been definitively broken by them; and a new problem has arisen: that of the relations between them and it. This evolution has been accomplished, in unequal degrees in different cases, and is today one of the chief elements of diversification in the universities. Princeton, for example, has no professional schools. Columbia has a very numerous series of them, among which appears even a school of journalism. We shall return to each class with some detail.

At the same time that the preceding transformation was being accomplished, an addition of a different order was being made to the old college— that of the Graduate Schools, and more particularly of the Graduate School of Arts and Sciences. The spirit of this addition was the introduction of original scientific research into the normal framework of the university.

Research had no place in the older college for the students, nor even for the professors. The impulse given by a few men, in the first rank of whom must be mentioned Louis Agassiz and Asa Gray, at Harvard, was the point of departure of this new era. Agassiz, who had been given a professorship, thanks to the organization of the Lawrence Scientific School, had founded in 1860, at Harvard, the Museum of Comparative Zoölogy, and had actively developed it through his voyages of exploration. He had grouped around him quite a nucleus of young men, to whom he had given the taste for original research. The palaeontologists likewise, a little later, Marsh at Yale, Cope at Philadelphia, as well as J. Leidy, attracted pupils. But scientific resources in America were insufficient for the pioneers in almost all branches of science, and the young men came to Europe for their apprenticeship.

England, in spite of the community of language, did not, at the time, offer them favorable scientific institutions. Cambridge and Oxford still remained confined to classical studies and to their old traditions. In France the Faculties were in the rudimentary state to which Napoleon I had reduced them. The great scientific men, like Pasteur, Claude Bernard, Sainte-Claire Deville, did not have laboratories in which they could have numerous collaborators. They understood the necessity for them, and asked for them insistently,[1] but without success, invoking, since 1867, in terms which have lost nothing of their worth, the example of Germany.

Youthful Americans were naturally drawn to Germany, because they found there all the necessary conditions for their apprenticeship, no examinations constituting, as with us, too numerous barriers; the possibility of winning easily the degree of Doctor of Philosophy, which they carried back with them as a palpable sanction of their work abroad; finally, well-furnished laboratories and seminars, in which the spirit of research was general, and the professors were devoted to their pupils. Already, about 1825, the Liebig

[1] See the "Report on Physiology" made by Claude Bernard for the Exposition of 1867, and Pasteur's requests to Napoleon III, in Vallery Radot, *La Vie de Pasteur*, pp. 204–206.

laboratory had begun to attract foreigners. We cannot too much regret that at the moment when Prussia, in 1811, founded the University of Berlin, and directed it toward original research, Napoleon I conceived the university faculties as mere bureaus for state diplomas. In the relations between America and Germany, the universities have been a factor of the first order. Germany has drawn from them not only important moral support, through the influence which she has temporarily exercised in a profound fashion on American mentality, but also, considerable material profit. It would be puerile to try to deny that she owes this result to the development of her laboratories and to the systematic direction of her universities toward original research.

During more than forty years, a good part of the most intellectual American youths, those who hoped to fill the chairs of new or enlarged universities, and who were in their time to shape the following generations, have gone to finish their education, and above all to begin research, in Germany. They received a profound impress there. At the beginning of the twentieth century, the vision of things scientific, in America, was through German ideas. Charles S. Minot, professor of embryology at the Harvard Medical School, expressed this fact in a very categorical manner, speaking of himself at the beginning of the opening lecture of the course he gave as Exchange Professor at the University of Berlin, in 1912. "Forty years ago," he said, "a young American, twenty years of age, decided to devote himself to science. He soon recognized that a young naturalist was far from finding the necessary facilities and support in America at that time. Therefore he resolved to come to Europe. He found in Germany teachers fired with sacred zeal, and laboratories, and so it happened that, through his German scientific education, he became an *intellectual subject of Germany.*" [1] In the American universities there is scarcely a professor belonging to approximately the same or the following generation, who has not worked in the German laboratories and who has not been profoundly influenced by the idea of the scientific supremacy of Germany. This has become a truism which is found endlessly expressed in the most diverse forms, in speeches, toasts, and otherwise.

We must be frank to recognize that this influence exercised by Germany, however excessive it may be, rested on solid bases. The Americans have learned much from Germany; they could bring back, for many sciences, models which they only had to adapt to their needs. That is past now. The apogee of German scientific influence had already passed before the war. The young American no longer needed, in general, to go to Europe to study. He had laboratories, libraries and guides at home. But the habit begun, the tradition spread, made many others take the same road as their elder brothers. The vitality of the German laboratories was thus assured for a notable part, by the foreign patronage which attended them, in particular by that of the Americans. Seeing Germany especially through science, the Americans had acquired tenacious illusions concerning its general mentality, which the war has dissipated, and which had completely disappeared before the United States came to direct intervention. The era of regular migrations to Berlin, Leipzig, or Heidelberg is doubtless closed for a long time.

[1] *Science*, December 6, 1912.

College remained the necessary foundation for students who wished to undertake really higher studies and research. The school of advanced studies was, then, purely and simply added; the students were men who had previously taken the Bachelor's degree, that is to say, graduates; and it received the name of the Graduate School of Arts and Sciences. It covers the field of our Faculties [1] of Letters and of Sciences, and extends the college in all its branches.

But not every college has such a school. It exists in scarcely more than thirty universities. It is best represented in those I mentioned in the preceding chapter.

The foregoing evolution, from college to university, has broken the unity of the former; and the relations of the parts, in the new organism, have not yet arrived at the state of equilibrium. There is a college crisis, which we see frequently denounced by the partisans of the tradition.

The classical college, in fact, with its four years of disinterested culture, preparing directly for no career, and holding the student till he is twenty-two, is too long a stage if one is to enter on professional studies afterward. Moreover, the professional schools which require a Bachelor's degree for entrance are exceptions, and indeed recent exceptions, even in such a case as that of medicine, where the Bachelor's degree is theoretically required. There again Harvard and Johns Hopkins have shown the way. It was to avoid the roundabout way through college that special engineering schools,

---

[1] I should remark here once for all, that the word "Faculty" is not absolutely equivalent to ours. It designates, in fact, almost exclusively the body of professors. The institution is called a "School," *e.g.*, school of medicine, school of law. In a college, the Faculty is the whole body of the instructors.

such for example as the Institute of Technology at Boston, were created — which takes its students at eighteen, exacting less knowledge than for entrance to Harvard, and brings them to their goal in four years. If the college kept rigorously its traditional four years, it would be deserted by many. It must therefore shorten its studies and combine them with professional courses. It must bring about a more complete inter-penetration of college and professional schools. This is the change which is being accomplished more and more, still meeting with a certain resistance.

The shortening of the college course can scarcely re-sult otherwise than to the detriment of the last two years. But, its defenders say very justly, those are the most essential for the training of the mind. The true solution of the problem would lie in an improvement of secondary education, which should bring back to the high school the present first two years of college, and would bring the student entering the university at eighteen to the state of maturity and knowledge which he does not reach today before twenty, when he becomes a junior. This view has been maintained by many uni-versity presidents and professors.

American secondary education is very short. It does not begin till fourteen, and covers only four years. Be-sides, the studies are less tyrannical than in France or Germany. The adolescent has much more leisure, which he can spend in games and sports. This pro-duces a much more vigorous youth. But from the intel-lectual point of view, there is an undeniable delay, and it surely seems that a better ordering of primary and secondary studies would resolve the difficulty at least in part.

The American university, in a general way, is still in a period of transition and of formation. The past persists, and remains its solid foundation; all that has been added to it and forms the superstructure, is heterogeneous, and the relations of the parts among themselves and with the whole have not yet assumed a character of definitive stability. The working out of this, is one of the chief differences between the several universities. In each it has resulted from particular circumstances and has taken place in a more or less special manner.

CHAPTER III

THE EXTERNAL ASPECT OF THE UNIVERSITY

The campus. Harvard, the Yard and the various additions. Columbia. Princeton. Berkeley. Cornell. Contrast with the French universities.

AFTER having looked at American universities and the prominent features of their historical development, in a general and abstract way, let us now approach them in their concrete reality, as they appear to us in their location and external appearance. A few examples will be the best means of giving an idea of them.

Let us go first to Harvard. Cambridge had remained until a short time ago, a peaceful town, with frame houses, each with its yard, in the midst of century-old trees. The yards are disappearing, and tall stone apartment houses, built close to one another, are little by little replacing the frame dwellings. The Gipsy Moth, imported from Europe — without the parasites which there limit its multiplication, having crossed to America at the same time — has propagated itself in a disastrous manner in New England, destroying the woods, and in particular killing many fine trees in Cambridge. Harvard today is no longer in a sylvan site. Little by little it has been surrounded by the less happy setting of the city.

The old Harvard of the College — which is generally called the campus in American universities, but which is usually designated here by the English equivalent, yard — is a large quadrilateral partly surrounded by walls and tall iron fences, partly by a plain wooden fence, which allows glimpses of its trees, over which many gray squirrels scamper, and of its lawns, in the midst of which rise the buildings or halls of the university. The latter are of brick, severe in aspect, the oldest without ornament, following the Puritan tradition; the old dormitories where the students live, the chapel, the administration building (University Hall); the President's house, rebuilt but a few years ago by the present President, Mr. A. L. Lowell; a group of buildings housing various departments of the university, Sever Hall, Emerson Hall, the school of architecture; finally, the monumental Widener Memorial Library, dedicated in June 1915. It is a city, with open spaces and well-ordered shade, but in which for a long time there has been no room for new buildings.

Thus, many years ago, Harvard began to expand. Memorial Hall, facing the Yard, a large building surmounted by a tower, was erected in memory of the Harvard men who fell on the battlefields of the Civil War. It was not thought that life should be excluded from this memorial monument. One of the wings is the large dining hall of the university, where one thousand students may dine together; the other is arranged as a theatre (Sanders Theatre), in which until recently the presentation of diplomas at the end of the year took place, and where from time to time dramatic performances are given. I saw there one of the farewell performances of a great English actor, Sir Forbes Robertson, playing Hamlet, in the simple setting of the time of Shakespeare. On such an occasion, the university receives its guests.

Beyond, in the still open part of the city, are dispersed along shady avenues a number of university buildings; laboratories, the Museum of Comparative Zoölogy (Agassiz Museum), and Museum of Ethnography (Peabody Museum), Law School, Divinity School; in another direction on the banks of the Charles River, the vast new dormitories which Harvard has just built in order to keep its freshmen together, and cement their comradeship by life in common, from the time of their entrance into the university.

One cannot help regretting that instead of developing somewhat by chance, outside of the Yard, Harvard could not have reserved for itself, when there was yet time, all the land which separated it from the river, and into which the city has now extended its narrow and ugly streets.

On the other side of Charles River, facing the freshman dormitories, a wide vacant space, Soldiers Field, is used for the manoeuvres of the Harvard regiment — they have been active for two years — and, on one side of it is the Stadium, an open air amphitheatre, on the tiers of which more than 25,000 spectators may be seated to view the games. On the banks of the Charles the boathouses complete this group devoted to physical exercise, which is so important in all the American universities.

But what remains crowded around the primitive nucleus is still only a fraction of Harvard. At a little distance in Cambridge are the buildings and dormitories of Radcliffe College, for women, distinct from Harvard, but affiliated with it. Crossing the Common, characteristic of all the old Puritan towns of New England, we reach the Observatory and the Botanical Gardens, with the building which shelters the Gray Herbarium, founded by Asa Gray.

In Boston, Harvard has its magnificent Medical School, at Longwood, in the hospital section. Rebuilt in 1907, it consists of five large marble buildings forming three sides of a rectangular court.

Finally, beyond Boston, at Forest Hills, Harvard has other dependencies; Bussey Institution, at first a school of agriculture, is today an institute of applied biology in which experimental heredity especially is studied.

Adjoining is the Arnold Arboretum, a magnificent park of 125 acres. Further away at Petersham, Harvard owns a forest of 1000 acres, which is a practical school of forestry. Besides these annexes in the neighborhood of Boston, there are more distant ramifications: a camp for students of mining engineering in Vermont, another camp, covering nearly 750 acres, in New Hampshire, where civil engineers serve their apprenticeship in the study of topography and in laying out railway lines; finally a biological station in the Bermudas.

Therefore Harvard is not a monumental but unexpanding building, confined in the unchangeable surroundings of a city. It failed, nevertheless, to grow in time to remain entirely concentrated.

Columbia College, stifled in old New York, emigrated exactly twenty years ago, upon becoming a university, to Morningside Heights, and is once more already shut in within the city. Built almost at one time, and after a unified plan, it has buildings of homogeneous style and some open spaces for building still remain. But when its prodigiously rapid growth is considered, it seems that it will soon be in straits again, and that it will perhaps think of finding a new location. Nevertheless it is at the time adapted as completely as possible to the life of a university in a great city. Each class of services has its own building and its separate entrance. The museum of archaeology is not near the chemical laboratory as at the Sorbonne. At the same time it has the benefits of concentration. One of its buildings, very suggestive in its present condition, has the form of a mighty semicircle, which ends abruptly at the second story, with a flat roof. This will become the foundation of a great amphitheatre, when the generosity of a donor permits the complete execution of the plan. And this visibly unfinished building seems to call for the donation. Upon entering it, you find in the basement a large pool where, at all seasons, the boys come to swim, and above which is a vast gymnasium. The main floor is a power-house, a central plant, which distributes heat, cold, compressed air and electricity to all parts of the university. All that, managed by expert engineers, assures to all departments the most modern services, while avoiding expensive duplications.[1]

At Chicago, that immense city, the university, founded in 1890, is also quite agglomerated, and does not yet lack room. It extends along a broad avenue, the Midway Plaisance, which connects two large parks. In 1914 it covered 90 acres and consisted of about 50 buildings, in an English Gothic style, very sumptuous as well as homogeneous, which recalls both Oxford and Cambridge. It has secured the ownership of the land in its

[1] At the Sorbonne, built at the same time as Columbia, the Faculty of Sciences alone has seventeen heating plants, but has no energy producing station.

neighborhood, bordering the avenue, and it can extend at its pleasure in the future. The University of Chicago is the one which in exterior has perhaps the best appearance and the amplest room, as an urban university.

The universities which have still remained outside the great cities, in the open country or in small towns, are more charming.

Such is Princeton, in New Jersey, scarcely two hours from New York. The town has only a few thousand inhabitants; it blends with the country in all directions, and seems to be but the necessary complement of the university. Along broad streets, lined with large trees, or widely and irregularly spaced on vast lawns, the sixty-five university buildings, laboratories, halls, and dormitories, seem scattered over a great park. Some date from the eighteenth century, and were witnesses or the seat of important events in the War of Independence. There was a battle at Princeton, and in one of the university halls, George Washington received the first ambassador accredited to the United States.

Princeton leaves with the foreigner passing through, above all an impression of luxury. Its Graduates' College, in the style of the great English colleges of Oxford and Cambridge, is particularly sumptuous. The students have numerous and elegant clubs. Out of the little stream an elongated lake has been made, Carnegie Lake, in order to allow canoeing and regattas. This seems to the traveler a Thelema's abbey for youth, and this impression cannot be entirely false, for Mr. Woodrow Wilson, who was its president before he entered the White House, made the following remarks, in a report which roused storms of protest. "We realized," he says, "that, for all its subtle charm and beguiling air of academic, Princeton, so far as her undergraduates were concerned, had come to be merely a delightful place of residence, where young men, for the most part happily occupied by other things, were made to perform certain academic tasks; that, although we demanded at times a certain part of the attention of our pupils for intellectual things, their life and consciousness were for the rest wholly unacademic and detached from the interests which in theory were the all-important interests of the place. For a great majority of them, residence here meant a happy life of comradeship and sport interrupted by the grind of perfunctory 'lessons' and examinations, to which they attended rather because of the fear of being cut off from this life than because they were seriously engaged in getting training which would fit their faculties and their spirits for the tasks of the world which they knew they must face after their happy freedom was over."[1]

I hasten to say that Princeton otherwise gives undeniable proof of being an important scientific centre, where investigators must enjoy a particularly calm and agreeable life. Works of the first order have come from its biological laboratories, and in particular, Professor W. B. Scott, who is one of the masters of the palaeontology of Mammals, has built up there, with materials discovered, worked up, and studied by him and his pupils, one of the finest and most valuable museums for this special branch, whose bearing on the study of the problem of evolution is considerable.

Properly speaking, the University of California is no longer situated in the country. The city of Berkeley is

[1] Princeton Alumni Weekly, 1907. Quoted from Slosson, *op. cit.*, pp. 79–80.

developing rapidly around it, but is spread out broadly in the midst of gardens. The university campus occupies a delightful site, on the slope of the hills, bathed by San Francisco Bay, exactly facing the Golden Gate, where, every evening, the setting sun plunges into the Pacific, framed by the silhouette of the mountains and of the great port. This campus is a vast park, in which the eucalyptus grow beside palms and numerous century-old live oaks with robust gnarled branches. The university of Berkeley, founded in 1868 by the union of a private college and a creation of the state of California, in execution of the Morrill Act, has had the happy fortune to have at its disposal an immense space. Its first laboratories were built of wood and are still standing, but as temporary buildings. A competition among architects was held, a few years ago, in which our compatriot, M. Bénard, was the winner, to plan the whole of the permanent buildings, and little by little the latter is going up, all covered with white marble. Already the administration building, California Hall, the college of agriculture, that of mines, and the library, have been finished, and others were in construction in 1916. In the centre, a replica of the Campanile of Venice has been erected, and on one of the slopes of the park, a Greek theatre, exactly reproduced, in which, thanks to the California climate, performances can be given in the open air before thousands of spectators. So the university city rises, little by little, without destroying nature.

Yet the vast campus at Berkeley only encloses a part of the university, the classical college, that of engineering and that of agriculture, as well as the scientific laboratories. In San Francisco, on the other side of the bay, which the ferries cross in twenty minutes, are the schools of law and medicine. This university has not had the too strict constraint of the old traditions of the college, and like the other state universities, it has developed broadly toward agriculture and the applied sciences. At the same time, gifts have furnished it great annexes for pure science, like the Lick Observatory on Mount Hamilton, and the biological station which Professor Ritter directs, at La Jolla, near San Diego, on the Mexican frontier.

I did not have a chance to visit Cornell University, at Ithaca, in New York state,[1] and I regret it, for, in a quite different landscape, it evokes the same happy ideas as Berkeley.

From M. Paul Marchal's book, which I have already had occasion to cite, I quote the following description, which gives at once a very lively and a very attractive impression of it.

"It spreads over a large wooded plateau," says M. Marchal, "bounded by cliffs which overlook the town and beautiful Lake Cayuga. Isolated by rocky gorges, through which narrowly confined torrents fall in cascades, it is accessible only by suspension bridges thrown from one wall to the other, and crossing above the gigantic tops of the century-old tsugas.

"This land, which measures not less than 1200 acres, is an immense stretch of verdure, woodland, and prairie, whose continuity is broken only by avenues and paths permitting approach to the various university buildings. A complete city rises there, whose buildings, isolated from one another, emerge from the midst of luxuriant foliage. First there are the many buildings

[1] The school of medicine of this university is at New York City.

in which are sumptuously installed the departments of the eight colleges and of the school of advanced studies, which compose the university. Of very diverse architectural types, often half-veiled beneath a mantle of climbing plants, they display the perspective of their gables and porticos along shady avenues, or are arranged in gigantic quadrilaterals, around carpets of verdure with trees in quincunxes. Farther on, in the charming setting of an English park, are grouped on a slope, and under the shade of large trees, the luxurious houses belonging to the different clubs or university associations (fraternities). Finally, the extreme northeast of the campus is occupied by dwellings for the president and professors of the university. They are grouped in a charming hamlet, composed of cottages scattered among trees and flower beds. Dominating the whole, rises the tall silhouette of the campanile, which thrice a day, in a sweet and joyous melody, sends forth the call of its chimes."

Such is the real setting in which Americans of today place their new universities. This civilization, which is above all urban, and whose cities are immense, nevertheless has not lost the feeling for nature. Are not students and professors incited to broad and living conceptions by always contemplating a wide horizon?

What a contrast with our stunted Faculties, rebuilt even recently in the centres of cities, and which no one dared to put outside them, in spite of the idea having been formulated. The Sorbonne, Darboux said very justly, is arranged like a trans-Atlantic mail liner. That is to say, for active life and needs which exact the broadest foresight and the greatest freedom of transformation, we have placed ourselves in the severest conditions of confinement. Thus the Sorbonne was not finished when already new departments which required a place in it could not be accommodated.

It is true that the American universities are, in themselves alone, cities whose numerous students suffice to populate and enliven them, and with university customs which are not ours. They can be sufficient to themselves, without being tributary to the great city. But whoever has contemplated the lawns of their grounds and the verdure of their trees, where the laboratories are hidden away; whoever has seen a numerous youth, full of the joy of living, passing through them, cannot but find all our university buildings terribly sinister, be they of the purest Louis XIII style, and cannot but pity those who have to study in them. One would be tempted to complete the republican motto on the pediments of our buildings with the line from Dante:

"Lasciate ogni speranza, voi ch'entrate."

CHAPTER IV

UNIVERSITY ADMINISTRATION
THE GOVERNMENT OF THE UNIVERSITY

Harvard, the Corporation and the Board of Overseers. Part played by the Alumni. Other universities. Trustees and Regents. The President. His powers and position.

WE must now see these great universities in action. We shall examine first the central organ which regulates and coördinates their activity, their administration. What precedes has given an idea of the multiplicity and complexity of their machinery. They have a large population of students, often running into the thousands, a working force of professors and instructors which reaches up to 700 or 800 persons, and they are at the same time, great estates with many buildings; they have funds of twenty and thirty millions of dollars, and an annual budget which often amounts to more than two million dollars of expenditures.

Here again, under the diversity of details due to the complete autonomy of each of them, in the present and in the past, we find at bottom a very great uniformity, which reflects the college tradition and the American mind in general. The present-day university is governed, in a word, as was the college, in spite of the transformation which has taken place. The tradition, and especially the spirit of the college, survive with perhaps excessive vigor. The result is, sometimes, as we shall see, very intense friction.

As the first example, let us again take Harvard, in which the friction in question has been reduced to the minimum, and is scarcely felt. In its broad lines Harvard has kept its seventeenth-century constitution, save that the representatives of the state have little by little been completely eliminated from its councils. The executive power is in the hands of the Corporation, which consists of the president, the treasurer and five members, the President and Fellows of Harvard College, and is self-perpetuating. The Corporation manages entirely the finances and property, chooses the president, appoints and recalls the professors, and grants the diplomas. Its power is without appeal, but nevertheless it is controlled by a sort of advisory council, provided with the right of veto, the Board of Overseers. In its present form, the latter has thirty members, chosen for six years, and renewed each year in groups of five, by election. This election takes place at the festivities at the end of the academic year, at Commencement, when the diplomas are presented.

All former graduates, the alumni, present at these festivities, have the right to vote. This board is therefore a direct emanation from the body of the alumni, who thereby exercise a general control over the progress of the university.

This participation of former students in the management of the university, totally unknown to us in France, is a heritage from the English tradition. It is a powerful bond between the institution and all those whose *Alma Mater* it has been; it makes of the university a really living and loved person, and not an abstract emanation from the state. It is a fundamental trait of the constitution of every American university, and it finds its place even in the state universities. "It is natural and proper," says Mr. Charles W. Eliot,[1] "to

---

[1] *Science*, December 15, 1905.

give some influence over the fortunes of a college or university to the body of its graduates, as soon as this body becomes large and strong."

At Harvard, however, the alumni exercise their influence only by way of control. They have no power over the composition of the Corporation, and the latter, at least at present, includes no member of the Faculty, scholar or professor, outside of the president; its members are Harvard men who have arrived at a high social position, business men, bankers, prominent citizens, like Mr. Robert Bacon, who was ambassador to France. The president alone therefore, represents the truly technical side; his fellows can help him especially by their business experience, in the financial management of the university.

The overseers, in fact, are also chiefly social notabilities; the share of the intellectuals is small, and many regret it. That expresses the fact that the dominant preoccupations of the body of the alumni are not of an intellectual order. They love profoundly their university, they interest themselves in its prosperity, and sustain it materially with a mighty generosity, but in the memories of youth which attach them to it, the intellectual side plays but a minor part.

The system with two bodies, which Harvard offers, is an exception. In general there is but one council, ordinarily called Board of Trustees or in the state universities, Board of Regents. In some cases this board is self-perpetuating; more often it is elected, at least in part, by the alumni. In the state universities it includes members *ex officio* such as the governor of the state, and members elected either by the state legislature, or directly by the people. In these universities, politics weighs more or less heavily upon their government. But it must be observed that this is not directly in the hands of the general government. There is always a council interposed, and as a consequence, large autonomy. Sometimes the board of trustees is very numerous, and in that case it delegates most of its powers to a commission of which the president is a member.

I cannot think of describing here in detail of varieties which different universities offer. There is, nevertheless, an interesting example, Cornell University, which is of a hybrid nature, a private university by its foundation, and a state institution in that it has received lands allotted to New York state under the Morrill Act, and because, moreover, it still receives other special subventions from that state. Its Board of Trustees is very composite. It includes fifteen self-perpetuating trustees, ten elected by the alumni (among these, a woman was elected in 1912), five designated by the governor of New York state, and ten members *ex officio*.[1]

The character common to all these variants is that the professors have no share in the constitution of the governing board, and that of all interests concerned, those of an intellectual and technical order are the only ones not directly represented in an assured manner. That is incontestably a defect, against which numerous voices are justly raised.

The council of trustees or regents governs, in a general way, the whole university, as formerly it governed

[1] The Governor and Lieutenant Governor of New York, the president of the legislature, the State Commissioner of Public Instruction, and State Commissioner of Agriculture, the President of the State Agricultural Society, the trustee of the Ithaca Public Library, the President of the University, and the eldest of the male descendants of Ezra Cornell.

the college. However, with the growing diversification and the enormous extension of the university with the special conditions under which each of its parts functions, certain of them must have more or less autonomy, and their own council, provided with greater or less powers. Here are some examples. A few years ago, Columbia University incorporated an institution till then distinct, Teachers' College, at once a normal school amd a school of applied arts, which is very large and by itself numbers about 2000 students. This college has kept its own board of trustees and governs itself. The Scripps Institution for Biological Research, a biological station established at La Jolla, near San Diego, at the southern extremity of California, is connected with the university at Berkeley. Founded with gifts specially restricted to it, it has its own council, which is in fact autonomous, whose decisions must merely be, in principle, confirmed by the regents of the university. You can imagine the multiplicity of degrees with which such autonomy may be invested, according to the circumstances and the flexibility which that possibility assures.

The board of trustees was an organization on the scale of the older colleges, in which the unity was absolute, and which had only a limited number of professors and students. It needs to be adapted to the scale of the new institutions and to their needs.[1] The technical incompetence and the excess of power of the trustees or regents are evidently a serious fault of the

---

[1] See on this subject, the projects of reform, in a very democratic spirit, suggested by Professor J. McK. Cattell, "University Control," *Science*, May 24-31, 1912, and the inquiry organized by him, the results of which the same journal has published.

present régime, and this fault has increased through over-growth, which is a peril to universities as well as to organisms. The problem is evidently to give a sufficient autonomy to the parts which are individualized, while maintaining a coördination in the whole.

In reality, there are correctives for the system, and I have seen one functioning at Harvard, which appeared very interesting to me, that is, what is called the visiting committees. The Board of Overseers, to accomplish its mission of control, names special commissions for each of the schools or institutions, or even for each of the departments which compose the university or the college. The members of these commissions are former students, chosen either for their competence or for their moral weight. I saw one of them accomplishing its mission with the absence of formalism and the gentlemanly spirit which impregnates the whole mind of this society. The members of the committee can gather from everyone suggestions and complaints of every nature, weigh them in their conscience as men not deformed by official life, and carry the echo of them to the overseers, themselves charged with the general control of the university. That is evidently a very flexible system, and one which grows out of the habit of self-government of the old protestant English communities.

The president of the university is the head and the working hand of the board of trustees. He carries out their decisions and proposes to them the measures he considers necessary. At least that is the usual case. There are a few universities in which the president is not a member of the board, but is merely answerable to it.

In reality power is concentrated in the hands of the president, who alone follows closely the life of the university, and is, in most circumstances — at least those which concern internal management — the only competent authority. His almost unavoidable policy is to induce the trustees or regents to leave to him the maximum of initiative and freedom.

The president of an American university has thus considerable power, and is only exceptionally subjected to effective control. In any case, the professorial body cannot take any action contrary to him. The president has at the same time a considerable and precise responsibility of direction. Is not the President of the United States himself invested with enormous powers? In every business, the man who is at the head, and who also is generally called the president, has almost absolute powers of management. There is room in America for energetic men who love action. They are not trammeled there. The seed can grow. The environment is propitious, for it does not lead, as with us, to the irresponsibility which destroys character.

On the president of the university rests the care of assuring the progress and material success of the institution, and as its material needs are always large, one of his principal functions is to find the goodwill to furnish the necessary resources. He must conciliate the legislature in state universities, or in private universities adroitly rouse the generosity of the alumni a task which is not always easy in spite of the loyalty of the alumni. Every alumnus, says a president, wants to see the college grow, up to the day when you turn to him.

As to the personnel, the president has almost discretionary powers. In most universities, the choice of professors, their nomination, promotion, and recall, are without appeal in the hands of the trustees, and in fact of the president. It is not astonishing that this régime sometimes gives occasion for very just discontent. Mr. J. McK. Cattell, whose democratic spirit is very much opposed to this function, says of it that it makes the president rather a boss than a leader. "In the academic jungle," he humorously says, "the president is my black beast." [1]

The president is the tyrant, good or bad. A good tyranny is a rule which has many advantages and the fact cannot be contested that certain American universities have owed a large growth and prosperity to having had at their head for a long time an active, enterprising president with broad and well-advised views. Mr. Charles W. Eliot, raised to the presidency of Harvard in 1869, at the age of thirty-five years, conducted the institution during forty years with a firm and sure hand, and under his leadership Harvard has been one of the chief guides in the evolution of American higher education. The first president of Johns Hopkins, Gilman, has played a rôle of the same order. The University of Chicago was opened in 1890, under the presidency of W. R. Harper, then thirty-six years old, and during the fourteen years he remained at its head it rose to the first rank.

At the present time the personality of the president is moreover particularly important. The American university, according to what we have already seen, is in a transitional phase between the college tradition and the spirit of real higher education and scientific

[1] *Science*, May 31, 1912, p. 845.

research. The equilibrium between these two tendencies, or its rupture for the profit of one or the other, is largely in the hands of the president.

The real paradox of the situation is that while having these extended powers and while thus administering the entire university from above, the president exercises authority directly in details, with scarcely any intermediate organs. Thus success in these functions is difficult in proportion to the power which they confer. The president of a new college in Oregon, Mr. W. T. Foster,[1] had the idea of visiting, at the beginning of his presidency, 105 colleges or universities, scattered over 29 states, and to inform himself upon the moral position of the president. In 51 cases he was able to form a clear opinion. There were 34, or two-thirds, in which the president distinctly gave dissatisfaction. It is evidently difficult for him to content everybody. He must be a scholar, says Mr. Foster, often a professor too (in many second-class institutions he continues to teach while president). He has the duty of watching the teaching of others. He must be a business man, and it is certainly less complex and more remunerative to direct a commercial business. He must find funds for the university, represent it, have happy relations with the alumni, students, and visiting strangers. He must be ready to speak at numerous meetings at any moment and on any subject; be able to guide the trustees through questions they are not acquainted with, to get along with the cliques in the faculty, to keep the professors patient, who are expecting promotion. The task is impossible, Mr. Foster concludes. It must be divided, while continuing to centralize responsibility.

[1] *Science*, May 2, 1913, p. 653.

The professorial staff suffers, in a general way, from the autocracy of the president,[1] except in those universities in which he knows how to use it with discretion, and in which, without being obliged by the regulations, he carefully consults, for example, the competent professors in regard to nominations to be made. But even with the best intentions, a man cannot understand equally well all needs and all tendencies. He will favor necessarily those which accord with his personal preferences.

Abuses of power and conflicts result from this situation, which are none the less unfortunate for being only the rather rare exceptions. Professors have been brutally dismissed from certain universities without even being permitted to defend themselves, simply for having expressed opinions displeasing to the president or trustees. There are evidently legitimate dismissals—although arousing the discontent of the interested parties — but the right of defense ought to be broadly assured, and the proof that, in more than one case, wrongs have been done, is that certain professors thus dismissed have been welcomed afterward by universities of the first order, such as Harvard. It has happened that facts of this kind have provoked resignations *en masse* and the exodus of most of the professors. In 1913,

[1] Here is a declaration coming certainly from a sincere conviction, but which seems to me very typical. The President of Vermont University, at his inauguration, declared to his faculty, "I should say to you, in perfect candor, at this time, in order that there may be no misunderstanding from the beginning, that I will not serve on a teaching body with any man who uses intoxicating liquors in any form whatsoever." (*Science*, October 13, 1911, p. 491.) He declared before elsewhere, that the use of beer and wine is degrading. One may judge from the categorical force of this declaration, to what lengths the president's power may extend and be exercised in practice.

at the University of the State of Utah, there were eighteen such resignations. These conflicts have ended by bringing about the formation of an association of college and university professors, which in the case of the University of Utah, sent a committee to the spot to make a regular investigation.

The omnipotence of the trustees and of the president, not counterbalanced by a control by the professors, is a subject of widespread uneasiness at the present time. In this regard, the American college has evolved from its English origins in a quite opposite direction to that of the English college itself. The latter is a monachal democracy, and the master is, among the fellows, only *primus inter pares*. A movement is undeniably appearing in the United States in favor of a more democratic reconstruction of the university.[1] This transformation, on the other hand, would have the inconveniences inherent in every democratic government, and we ought in truth to admit that the bodies actually administering the universities are in a very general way animated with perfect disinterestedness and inspired by an ardent will to assure their prosperity and success. To the trustees, especially in the old institutions, the university is a living and loved person, and not a cold administrative machine. The presidents too, have the highest idea of their task, and unreservedly consecrate to the development of their university the personal strength of character and energy for which they were chosen.

---

[1] See J. McK. Cattell, " University Control," *loc. cit.*

CHAPTER V

THE PROFESSORS

General conditions of the career. Moral and material *desiderata*. Excessive burden of teaching. Insufficient participation in the management. Precarious guaranties. Advances in the career. Salary. Retirement pensions. The Carnegie Foundation.

OF all the elements of which a university is constituted, the professorial staff is evidently the most essential. At times, in some countries, it has a tendency to believe that it itself is the only one of which it is necessary to take account, but this is evidently excessive. Still it remains none the less true that on the worth of the individuals who compose it, depends all the intellectual strength of the institution. Therefore the conditions of the recruiting and of the career of the professors have a great importance for the evolution of the university and for its scientific productivity. On the other hand, the professors represent a collectivity whose interests are distinct at once from those of the administration, which we have just studied, and from those of the students and alumni, which we shall consider hereafter.

If we except a few privileged universities, the material and moral position of the professors in the United States is modest. "The young American," says Mr. Charles W. Eliot, "who chooses the university career, must abandon all prospect of wealth and of luxury which a fortune alone can procure. What he can reasonably hope for is an assured income, a stable position, long vacations, the satisfaction of intellectual tastes, good comradeships in study, teaching or research, large resources in books, and an honorable but simple way of living." Still, that is a picture drawn by an administrator, inclined to optimism in the matter, on principle.

American professors make known today desires of two sorts, some moral, others material. We shall examine them in succession.

The first of these *desiderata* relates to the excessive burden of teaching. The essential function of higher education is research. It is not a question of sacrificing teaching to it, but the professors of a university must be left freedom of mind and time enough to undertake and conduct research successfully. However, in America, they have almost daily lectures. The majority of these doubtless do not call for much preparation. But meanwhile the professors have too many committee meetings and cares of an administrative order, and they must occupy themselves too much with the students individually. That is not the right pace to set for advanced teaching.[1] This is due to the college spirit, and to the unpreparedness of the students who come to it. But though the fact be explained, it remains, none the less.

The second desire of the teaching staff is included in the preceding chapter: it is the counterpart of the president's situation. In most universities the professors demand a regular share in the government of the

---

[1] Our American colleagues have, however, from the point of view of vacations, the very enviable privilege, every seven years, of being able to take a half or even a whole year of leave, which they call a sabbatical year. They receive full pay for six months, or half-pay if they are absent an entire year.

institution, and in the mechanism assuring the recruiting of its personnel, as well as greater security in the positions gained. The faculties deliberate frequently and long, but often on petty questions of detail. All that really concerns the general progress of the university remains outside of these deliberations.

As we have seen, the professors have no normal representation in the directing council. The influence of business men is valuable in order to administer skilfully the funds of the university, and that of the alumni is equally favorable. But neither the one nor the other is a guarantee for the general intellectual interests nor for the special interests of the professors.

Especially if it is a question of choosing a specialist to provide certain instruction, it is a principle recognized everywhere, although often insufficiently applied, that the advice of competent persons is one of the primordial elements of the choice to be made. Yet the college professors, the faculty, have no official and legal share in these appointments, which are the act of the president and trustees alone. In reality, broad-minded and well-advised presidents consult competent persons in their faculty, but only out of courtesy and at their good pleasure. There is very just complaint at that, formulated many times in late years, and the justice of which is recognized, besides, by many university presidents.

Without doubt it will come about before long, following the characteristic and fortunate principle of the English mind, that a reform must be proclaimed by law when it has already been consecrated in practice.

If the professors have no regular part in their recruiting, they also lack precise guarantees of the possession of their positions. That is due to a general trait of

American customs, which has advantages for society. There are no fully assured positions in which one can go to sleep in security and inaction, at the expense of the interests over which one is to keep watch. The plague of officialism is thus avoided. Everyone must constantly justify his function by real activity.

Most chairs are given in a temporary manner. The instructors are appointed annually; the assistant professors for short periods, most often for three years. The associate and full professors are appointed without limit of time, but without guaranty; during good behavior, or at the pleasure of the trustees, say many contracts. The administration thus has a weapon in its hands, which it can use at almost any instant against the professors.

It uses it in fact only in very rare cases, but it is none the less a redoubtable menace, and one against which the professors are at present without recourse. They desire irremovability, life-tenure of their chairs, at least in the higher grades, when they have been tried; it concerns their security and their dignity.

Above all they ask a regular procedure by which they can controvert the complaints which are lodged against them. They totally lack guarantees which higher teaching possesses in other countries. Accordingly from time to time, discussions arise, like that which occurred in 1913 at Philadelphia, following the dismissal of an assistant professor of political economy. It seemed that this dismissal was due to certain opinions expressed by him in public lectures given outside the university. The tone of the discussion which followed, in the newspapers, is especially characteristic, the fact itself being difficult to appreciate here. The *Public Ledger*, one of

the most important and weightiest newspapers of Philadelphia, summed up the affair by declaring, "The public
has every right to know whether its greatest institution of learning is free to seek the truth and to proclaim
it without fear, or whether it is constrained to keep
silent every opinion in political or economic matters,
which is not momentarily to the taste of the Trustees."

This affair [1] ended, moreover, in causing a modification of the statutes or by-laws of the university, by the
trustees, in a direction which recognizes that the claims
of the professors were right. A right of consultation
concerning appointments, of permanent appointments
as full professors, and appearance before a board of
their peers before every dismissal, have in fact been
granted to the professors in the University of Pennsylvania.

There is undoubtedly a very general movement of
opinion in the universities on these questions at present, which seems on the point of leading to important
reforms. The formation of an association of professors,
to discuss them and to bring them about, is a characteristic symptom. The uneasiness scarcely exists in the
most powerful universities, which are at the same time
the most liberal in regard to their personnel, as is the
case with Harvard and a few others.[2]

[1] Cf. *Science*, 1915, no. 2.

[2] In fact, these questions so to speak do not exist at Harvard. Tradition
has established in that old university, more than anywhere else, at once
among administrative officers and those under them, a sort of coöperative
spirit which, till now at least, has kept them out.

Most of the professors do not even wish for a more direct participation
in administration and in appointments, fearing that it would introduce intrigues which would be prejudicial to the spirit of trustful comradeship and
cordiality which reigns in the teaching staff. They rely on the corporation
the more willingly because in their eyes it incarnates the Harvard spirit

An interesting general study of these questions will be found in the articles published in 1912, in *Science*, by Mr. J. McK. Cattell, under the title "University Control." He prepared a complete plan of reforms, on the subject of which he conducted an inquiry among the scientific professors of the various universities. The intention of the reform he outlined was the democratization of the university organization, the reduction of the president's powers, and especially the subdivision of the university, which has become a giant, into smaller homogeneous units, as largely autonomous as the harmonious functioning of the whole would permit. I cannot enter into the detail of these propositions and of this inquiry here. The 229 replies which Mr. Cattell received to his questionnaire were rather divergent, as is natural in so complex a problem. He analyzes the reasons for which on the whole they reflect the general opinion very exactly. The great majority were clearly favorable to an extended reform, and 184, or approximately two-thirds, adopted the proposed plan in its broad lines. The universities in which the govern-

and devotion to the institution. The professors, they note, do not form so homogeneous a body. A notable part are of origin foreign to Harvard, and they wisely maintain that it should be so, in order to avoid the danger of inbreeding, while assuring constantly the renewal of ideas by the introduction of outside elements. Harvard reflects well the old English spirit, by which things draw their force from the consecration of use rather than from the written letter. Thus it is that for appointments of professors there is not even a written contract, and yet in fact irremovability is complete. Before all else, they count on mutual loyalty. Similarly, although the consultation with professors is not written in the by-laws, it takes place in fact. The general principle is to keep the maximum of flexibility in all things. So to speak, there are no permanent chairs. Vacancies are filled to meet the needs which the present state of ideas and of the sciences suggests, and not by tenaciously keeping a branch of instruction because it existed yesterday.

ment is the most liberal were those where the least changes were desired. Those in which the autocracy of the president was, on the contrary, the most effective, were the most in favor of reform. Some administrative officers, almost alone, favored the *status quo*.

Such are the complaints of a moral order which the professors formulate. The general organization of the American university is certainly behind that of the other great scientific nations, as to the independence of the teaching staff.

It is no less interesting to examine the financial situation of the professors. First let us follow them through the various phases of their career.

The Bachelor of Arts who is looking forward to a professorship, remains at the university as a graduate after his four years of college, and spends three years in becoming a Doctor of Philosophy, Ph.D., a title which he obtains when about twenty-seven or twenty-eight years old. If he is not rich, he has been dogged by physical necessities during this period. These necessities are often more or less alleviated by scholarships or fellowships provided by private foundations and in compensation for which he has often already taken part, as assistant, in teaching. These subsidies sometimes have the form of traveling fellowships. They have sometimes been criticized as tempting mediocre individuals, who without them would have been eliminated by selection. But it suffices in order to justify them, to show that some of the most noteworthy men have been able, thanks to them, to pass the difficult period. That is what Mr. E. B. Wilson remarked concerning himself.

About the time of his doctorate, the future professor arrives at his first regular functions, on being appointed instructor. The instructor is the equivalent of the *préparateur* or *chef de travaux pratiques* of our faculties of sciences, but he exists in every branch of teaching and not only in the case of the experimental sciences. He is charged with following closely the studies of a group of students. That is an excellent idea; the organization of a good corps of instructors is the best way of assuring the regularity and solidity of studies for the students. The instructor is in general reappointed annually.

The following stage is that of assistant professor,[1] which can be compared to that of our *maîtrises de conférences*. The assistant professor is generally appointed for limited periods of time, most often for three years, which may be renewed. From the intellectual point of view, he is completely free in his teaching.

After a rather long time passed as assistant professor — automatically at the end of eight or ten years in certain privileged universities, like Harvard — or, more often according to the circumstances and the vacancies, he becomes a full professor, that is to say, *professeur titulaire*, or associate professor. This last grade answers very well to our *professorat-adjoint*, but in many universities it is a sort of shelving process for those who

[1] Mr. G. Marx published in *Science*, May 14, 1909, March 18–25, 1910, an investigation of the professor's career and especially of assistant professors. The result of it is that the average age in this rank is thirty-six years. The average age at appointment is thirty-one. For 120 persons who were investigated, the average duration of their studies had been seven years. Sixty-five per cent of them had had scholarships, and 45 per cent of those in this latter class had not been able, nevertheless, to finish their studies without contracting debts which burden their finances for some time.

have hardly any chance of reaching the full professorship.

This hierarchy is very uniform in the various institutions, but beneath the unity of names the real situations are very different, materially and morally.

It will be noted that at Harvard, which in many respects can be regarded as the standard university, advancement is automatic, or at least largely independent of circumstances. That might produce what in France is called *titularisation personnelle*. But we must note, in justice to this advancement by the mere passage of time, undoubtedly deadly to an indispensable selection, that there are two capital correctives, which really maintain an efficacious selection. The first is the grade of associate professor, as has just been explained. The second is the fact that only the men who have a certain worth tend to remain at Harvard to round out their whole career. The inferior are brought by the force of things to emigrate sooner or later into the less important universities, or even abandon the calling in the first years.

To what in salary do the various degrees of the hierarchy correspond? If we represent the average pay of full professors by 100, that of the other grades is represented by the following figures,[1] in a few universities which I have chosen as examples.

	Harvard	Cornell	Stanford	Wisconsin	California
Instructor	23.7	29.1	33.1	38	...
Assistant professor	61.6	54.7	45.8	59	49.4
Associate professor	81.6	...	63.4	75	68.8

The instructor in the large universities begins with a salary of $1000 to $1200 and receives annual increases

[1] *Science*, May 14, 1909.

of $100 to $200 up to a maximum of $1600 (Harvard)
or $2000 (Columbia).

Salary	Instruc-tors	Assistant and Associate Professors	Full Professors	Total Number	Per Cent of Total	Comparative Percentage for the Personnel of the Public High Schools, in 1908
Less than $750..	51	....	....	51	1.2	30.5
$750–$1249.....	911	74	12	997	23.4	44.3
1250– 1749.....	386	447	147	980	23.0	13.0
1750– 2249.....	29	483	227	739	17.3	7.1
2250– 2749.....	3	194	266	463	10.9	3.1
2750– 3249.....	....	76	286	362	8.5	1.6
3250– 3749.....	....	17	205	222	5.2	0.5
3750– 4249.....	....	9	194	203	4.8	....
4250– 4749.....	....	....	67	67	1.6	....
4750– 5249.....	....	....	95	95	2.2	....
5250– 5749.....	....	....	40	40	0.9	....
5750– 5249.....	....	....	25	25	0.6	....
More than $6250	....	....	18	18	0.4	....
Total..........	1,380	1,300	1,582	4,262	100.0	100.1

The salary of an assistant professor varies, under the
same conditions, from $1800 to $3500; that of full
professors from $3000 to $5000.[1] Harvard and Columbia are, in a general way, the universities in which the
salaries are the highest and in which the personal situation is the safest.[2] In order to judge the financial
situation of all the professors, I reproduce the statistics[3]

[1] The normal pay of full professors at Harvard varies from $4000 to
$5000, by increases of $500 every five years.

[2] The percentage to the total of the salaries to the total budget of expenses varies from 37 per cent (University of Missouri) to 75 per cent
(Columbia, Princeton, Pennsylvania, New York University). The cost,
calculated per capita of the students, varies from $100 (University of Syra-
cuse) to $475 (Harvard, Johns Hopkins).

[3] *Science*, June 12, 1914.

condensed in the table on p. 59, and borrowed from a work executed by the Carnegie Foundation for the Advancement of Teaching. It deals with 61 colleges of varying importance and relates to the year 1912–13.

As another indication, taken from the same source, concerning 201 full professors of Harvard and Columbia:

4	receive less than	$3000	
12	" from	3000 to $4000	
35	"	4000	
30	"	4500	
44	"	5000	
38	" from	5000 to 5500	
21	"	6000	
17	" more than	6000	

The preceding figures, converted into francs, give much higher numbers than the French salaries. But in order to judge them, we must of course put them in the surrounding conditions of life. On the whole, the situation of professors of American universities is collectively notably better than that of their French colleagues, yet is not, for all that, more than mediocre. The pay is not at all in proportion to the severity of the previous selection, nor to the social function performed; and it furnishes the family, in the setting of American life, only a very restricted budget. Consideration for the profession by the masses, who judge according to the salary, is very slight. The university profession has not in general, the moral and worldly situation which it deserves in a rich democracy like that of the United States, and which it ought to have in order to retain a high class of men.[1]

[1] These considerations of course apply to France, where also the university career is too mediocre financially and too uncertain. They apply, moreover, to almost all countries. In Germany, by the confession of the most

The professors complain that the enormous development of the universities and colleges in the past thirty years has been made at their expense. The increase of their pay, contrary to the case of most professions, has not even followed that of the cost of living. Besides, the ratio of the number of full professors to that of the students and of the professors of lower grades has been constantly on the decrease. Access to the best positions thus becomes more difficult and more tardy, while the quality of the personnel has been improved. Many instructors of today, they say, are as good as professors of yesterday. Mr. G Marx [1] sees the reason for this disproportion in the fact that the managing councils, the boards of trustees, have too often sacrificed the interests of the teaching staff to the development of the exterior, in order to attract and maintain the patronage, by exaggerated increase in the number of branches of instruction, construction of sumptuous new buildings, exaggerated luxury of all the university

esteemed professors, it seems that the generation which is coming into its university chairs is not as worthy as the older, because the finest of the youth have recently been too much attracted by the development of industry. German universities have, moreover, owed a part of their vitality to the constant afflux of foreigners to its chairs. This afflux was explained, as we must recognize, in part by the strength of their organization and of their professorial staff, but also in part by the superstition of the entire world regarding the virtues of German things. The academic career itself is, in the generality of cases, very mediocre in Germany from the point of view of money. But it attracted, on the one hand by reason of the consideration which it enjoyed, and on the other hand because in almost all specialties there were some chairs bringing in a great deal. It was the big prize which each one, at the beginning, hoped to win, and which made them take the lottery tickets, that is to say, which determined them to enter upon the career. With us, on the contrary, the big prizes do not exist. There is no incentive to activity. Good and bad are recompensed about equally.

[1] *Science*, May 14, 1909.

life. They have almost always preferred to satisfy needs of this kind to the detriment of the personnel. These faults are of course greatest in institutions of the second and third class because of the spirit of bigness and the ambition to follow, at any price, the example of the great.

In order to complete this picture of the material and moral position of the professors, I will say a word about the end of their career and about the question of pensions. Aside from certain large universities, it seems that until a few years ago, nothing was provided on that score. The professors taught as long as they could, or as long as the trustees found them in a state of good behavior. It was left for them to take, by insurance, the necessary measures of foresight, and certain people are still of opinion that that was a good system. In these Anglo-Saxon countries, the individual is used to counting only on himself. Fifteen years ago, Mr. Andrew Carnegie, desirous "of serving the cause of higher education by improving the teaching profession and augmenting its dignity," devoted a part of his fortune to the creation of a system of retirement pensions. On the one hand, he instituted the Carnegie Corporation of New York, for "advancement and diffusion of knowledge and understanding," endowed with 125 million dollars, and in 1905 the Carnegie Foundation for the Advancement of Teaching. This latter, at the end of 1913, had been endowed by him with 15 million dollars, and under the administration of a board of trustees, composed mainly of university presidents, it was charged with organizing a system of pensions in the universities and colleges. Mr. Carnegie excluded, however, at the beginning, the state universities, thinking that the state must do what is needful for them; and all the sectarian institutions. Those only which are undenominational were called to benefit from the Foundation. It is admirable to see an individual propose to himself the realization of a work of such breadth.

The rules adopted by the Carnegie Foundation recognized the right to a pension at sixty-five years of age, and fifteen years of teaching with at least the grade of assistant professor, or indeed after twenty-five years of teaching thus defined, without condition of age, or finally indeed in case of infirmity. The amount of the pension is based on the pay of the last five years, with a maximum of $3000. It can reach as much as 90 per cent of that pay, when it does not go over $1600. Widows have a right to half of their husband's pension. All this is applicable only to persons for whom teaching is the essential profession, and not, for example, to physicians and engineers, for whom it is only an accessory resource.

The Carnegie Foundation has been working now for ten years, but it is not certain that it will not meet with serious difficulties. It seems that the pensions asked for require, and especially will require, an amount greater than the provisions. The managers of the Foundation announce today that the pension can be morally claimed only by professors whose forces have become enfeebled, against which protests are raised justified by the promises made. Mr. Cattell had already in 1909 formulated certain serious objections and expressed the opinion that the intervention of an individual should not excuse the institutions themselves from assuring the lot of their staff. The Foundation should improve on what the universities should have done, following their strict duty, by completing, for example up to the total of the salary, the fraction which the university had provided. According to him, the defect of the system is, that it has its maximum of efficacy in the cases in which the Foundation was the least necessary. On the other hand he fears that its existence will contribute to cause the resignation of professors against their will, a procedure which the university could not previously have brought to pass. This is not the place to discuss this question fully. But it seemed to me interesting to note its elements, especially because the conditions laid down are so far from our bureaucratic customs.

CHAPTER VI

THE STUDENTS AND THE INSTRUCTION

The classical college (undergraduate). Admission. Organization of studies. Departments. Coördination of courses. Examinations and graduation. College life. Social and collective life. The dormitories. Clubs and fraternities. Sports and athletics. Various associations, dramatic societies. The general results of college studies.

AFTER the administration and the professors, the students. With regard to them, we must take up one by one each of the parts which we have distinguished in the university: in the first place, the college. We have seen that it is the fundamental part, historically and actually. Many institutions are limited to the college alone; in most of the large universities it is numerically the predominant part. Of 5000 Harvard students, 3000 belong to the college. It is the college which still impresses on the university its characteristic traits. It is more or less distinct; in recent institutions it may not perhaps exist explicitly, yet its spirit persists and is on the whole constant. I will try to give an idea of it, especially as it is at Harvard, where I was able to observe it *de visu*.

And first, how is the college recruited?

The student enters college at about the age of eighteen years, after leaving the high school, or the secondary school where he has remained, usually four years, from the age of fourteen to eighteen. Normally studies continue four years, and the student is called successively freshman, sophomore, junior, senior. As has already been said, he enters with an intellectual training little advanced, and a not very homogeneous stock of knowledge. In fact, the elective system which had been introduced into the college has been carried over into secondary instruction in large measure. Each one tends to take secondary studies as suits his tastes, or rather, his whim. That is, perhaps, a general aspect of contemporary American mentality, in matters of education, and is certainly connected with the prosperity and absolute safety of this people, as well as with the ease with which it has been able hitherto to move over an immense territory of virgin riches. They try to compel the child as little as possible, to present life to it under its most smiling form, to spare it opposition, to make work appear to it under the form of pleasure rather than of duty. This is very striking, even though one lives only a little in the intimacy of a family. By virtue of this tendency, they treat the schoolboy too much like a student, to the detriment of healthy intellectual discipline. So he often arrives, after leaving the high school, with considerable deficiencies even in the knowledge of English.

A certain number of colleges admit students on mere presentation of high school certificates, which show that they have followed a regular course of secondary studies, and enumerate the subjects they have studied. Sometimes this certificate is only recognized when coming from qualified high schools, that is, from those on a special list, kept to date, according to results which freshmen of preceding years produce, and which permit them to estimate the schools from which they come. But many large universities, especially in the East, and this is the case with Harvard, admit their students only after a special entrance examination, which they arrange for each year, and which comprehends obligatory subjects and optional subjects.[1] Thus every freshman enters with a certain specialization, and a record is kept of his previous studies.[2]

Thus Harvard receives at present 600 or 700 freshmen a year. They constitute a class, which is designated by the year of graduation, that is, of the year when it will be the senior class. The freshmen who entered in 1913, for example, form the class of '17. Outside of these regular students, there are some who are admitted on exceptional conditions, and are called special students, or out-of-course, or unclassified.

The university lectures are not public. Only the regular students, specially enrolled in each of them, are admitted.

Harvard draws its students from very varied social classes, and from all over the United States. The university consciously makes an effort to be a unifying factor in the country.

We have brought the student to college: how are his studies regulated? He is in a general way much guided and much watched. Each one decides at the beginning the program of studies he will choose, following the elective principle, under the direction of a professor [3]

[1] Here is one of the combinations of subjects: 1, English, 2, Latin (or French or German or Spanish), 3, elementary mathematics (or physics or chemistry), 4, a fourth subject taken from those in the first three groups which had not been chosen.

[2] The students entering with part of their studies finished, coming from another college, are admitted with credit for the studies already completed by them, and for the time which they have devoted to them. The various studies are carefully scaled in value during the four years, and each study has a definite value.

[3] At Harvard each professor takes charge of four students each year. For them it is a chance, from the beginning of their course, to form friendly relations with him.

designated by a permanent Committee on the Choice of Electives. Thus a plan of studies is traced for him, adapted to the career which he counts on undertaking. During the first two years, certain subjects are obligatory, like English composition, and a minimum number of courses must belong to the same group of studies. It is easy to see that this system may be a burden for the professors, whom it obliges, outside of their lectures, to do very considerable administrative work, in guiding and following the students.

The courses of instruction are extremely numerous and all those which relate to one science or to a group of related sciences, constitute a department. Each department has either a head, who regulates all the instruction within it, or, in more democratic fashion, which is that of Harvard, a chairman, a sort of president-secretary, charged simply with the coördination of the work, and with the relations between the department and the central organs of the university.

These departments, which are enumerated here, are divided at Harvard into four large groups:

1. Languages, Fine Arts, and Music: Semitic Languages and History, Indic Philology, Classical Languages, English, Germanic Languages and Literatures, French and Romance Languages and Literatures, Comparative Literature, Fine Arts, Music. In addition, Egyptology, Slavic languages.

2. Natural Sciences: Physics, Chemistry, Engineering Sciences, Botany, Zoölogy, Geology and Geography, Mineralogy and Petrography, Astronomy, Hygiene and Public Health, History of Sciences.

3. History and Social and Political Sciences: History (with numerous subdivisions), Government (Constitu-

tions, General Legislation, International Law and Diplomacy), Economic Sciences, Education, Anthropology.

4. Philosophy and Mathematics; Philosophy and Psychology, Social Ethics, Mathematics.[1]

You see what a variety of subjects the college carries. It aims to furnish the possibility of complete general culture, including even the general parts of law and of the economic sciences. It must be added that in every department the courses are very numerous. As examples, I note in the Harvard catalogue for 1915–16, 27 distinct courses in the division of Semitic Languages, 6 in Egyptology, 22 in French, 9 in Italian, 8 in Spanish, 4 in Celtic, 8 in Slavic Languages, about thirty in Comparative Literature, 22 courses in Physics, more than 30 in Chemistry, 16 in Zoölogy, about twenty courses in Pedagogy (Education), 14 in Anthropology, divided over a cycle of two years. The variety is no less, if one opens the Register of the University of Ohio or Columbia, or of Cornell. At Cornell, where Entomology has had a very great development, in view of its applications to agriculture, it is represented by more than 20 courses.[2]

These courses are of different levels. At Harvard each one is classified by one of the three following characterizations: 1. Primarily for undergraduates, these are the fundamental and elementary courses; 2. For undergraduates and graduates, these are higher courses to

[1] This enumeration characterizes an institution very much impregnated with the classical education. In the state universities, of a much more narrowly utilitarian spirit, the classics hold, in a general way, a lesser place, and on the contrary, practical teaching, so-called, has a more or less preponderant place.

[2] See P. Marchal, *loc. cit.*, p. 264.

which students are not admitted until they have taken the corresponding elementary courses; 3. Primarily for graduates, these are advanced and specialized courses, intended especially for graduates, to which college students may be admitted in their last years, if they have the necessary knowledge.

Such a system lends itself to an almost unlimited diversification of studies. Most of these are half-year courses,[1] and comprehend two or three lessons, of an hour, or more exactly, fifty-three minutes, a week. In experimental sciences, one lesson is generally transformed into a laboratory period. Each course forms a class, in which assiduity is controlled; the work is watched. At the end of the half-year the student undergoes a written examination on each course, in which he usually answers ten questions. This number of questions tests him on the various parts of the course, but in a rather superficial fashion.

To obtain the Bachelor's degree, the student must pass in a satisfactory manner thirty-two examinations of this kind, spread over four years, or eight half-years. That leads him to take normally four courses a half-year.

It is thus clear what the college student's system of studies is during the four years which he spends there. The extreme flexibility of it must be remembered. Each one can push his studies in the direction which interests him. As for the value of the result, it depends on the ardor of the student in his work. We must have no

---

[1] The first half-year opens in the last week of September and lasts till the end of January; the second goes from February to about the beginning of June. The academic year ends between June 20 and June 25 with the Commencement festivities.

illusions about it; with the average, it is not very great. Studies are only one of the elements of college life, and for many they are not the chief element. Good scholars are not the glory of their class. The very diversity of studies tends to render them somewhat superficial. The Harvard studies and Bachelor's degree, under these conditions, are something hybrid between our studies in the *lycée* and our Faculties of Letters, or of Sciences, or of Law, put together. An A.B. graduate means more than one of our *bacheliers*. His age of twenty-two years gives him more maturity. There has been a character of freedom in his studies which is perhaps the true element of higher education. He has thus been able to push them in one direction, in which he will have acquired very profound knowledge.[1] To sum it up, the results will vary enormously according to individuals. They may be excellent.

What is true of Harvard is true of the other universities. The Harvard Bachelor's degree is undoubtedly among the best. Those of the 600 universities and colleges spread over a long scale of values. In a general way, what seems to be the greatest fault is the lack of solid training of the mind by secondary teaching. As Mr. Woodrow Wilson said, we must not confound education and information. In the American system, there

[1] Thus I had occasion to see, at Harvard, some very good students of biology. They had brought a taste for it to the university, and from their college years had been able to develop it. The biological sciences are among those for which the conditions are better in America than in France. Natural history has a rather large place in secondary and even primary instruction, and above all, it is taught by keeping closer contact with nature. Besides, nature is much richer and less deformed by civilization, even in the outskirts of many cities. The liking for excursions and camping is also a factor which opens up these callings in youth.

is, from the intellectual point of view, too much information and too little education.

One would have only a very incomplete idea of the
American college or university if one looked at it from
the purely intellectual point of view. The life, properly
speaking, of the student, especially its social aspect, is
an essential element of it, not only in the psychology of
the student, but in the thought of many educators.
The freshman is still a boy; the task of the college is
to make a man of him. The training of character has
an importance of the first order. It is a question of
learning, while living and learning to live. Our purely
intellectual university environment seems inhuman to
the Americans, according to the very expression of Mr.
Barrett Wendell, who has nevertheless seen it with so
much sympathy. "The object of the college," said Mr.
Lowell in 1909, on assuming the presidency of Harvard,
"is not to produce hermits, each imprisoned in his intellectual cell, but men, adapted to take their place in
the community, and to live with their fellow-workers."
He says further, "The college produces liberty of
thought, breadth of views, training of the civic spirit."
This, more than a high intellectual training, is the impress which it leaves on the majority of its students.

This education is based on the life in common and
the development of sociability. In the old semi-
ecclesiastical college, the life and the studies were common to all. The spirit of the English colleges of Oxford
and Cambridge, with the production of the English
gentleman as its ideal, was continued. The diversification of studies and the multiplication of the number of
the students have destroyed this unity. Many uni-

versities, and Harvard in particular, nevertheless force themselves to maintain it. With this intention, Harvard has built, in these last years, along the Charles River, four large dormitories, in which the students live. Each man has his individual room or apartment, or groups of two, three, or four share an apartment, giving to all the bathroom, which seems a luxury to us.[1] A refectory brings together all the youths in each one of these buildings, and in a large and comfortable hall they can live together in leisure hours, read papers and magazines and play music. From their entrance into the university, they are thus turned away from the solitary and individualized life which is that of our student youth, and are led toward a different psychology. A few older students, the proctors, are invested with a certain authority and charged with maintaining order and good behavior.

In the following years, a part of the students still live in dormitories. Others live in groups in houses,

[1] It seems interesting to me to give an idea of the student's budget. The tuition is $200 a year, for the lectures, to which must be added laboratory fees, for each course in the experimental sciences ($2 to $5 or $10 a half-year course). The state universities admit free, so far as tuition is concerned, students of the state to which they belong; and others at a very low tuition.

Rooms in the Harvard dormitories range from $30 to $350 for the academic year, according to luxury. The student can board in the university dining halls for from $5 or $6 a week up. The university has organized co-operative societies, where the students can buy their books and all sorts of merchandise very advantageously.

The expenses of a student for the nine months of the academic year seem to be at least $600.

At Harvard, there are about 300 scholarships for the college, whose value ranges from $75 to $300. About twenty, however, are higher than this figure, and one even reaches $700. They are given for merit. In the universities, moreover, it is not rare for students to accept, in order to pay their expenses, jobs which among us would be considered servile, but which in America do not lower them at all in the minds of their comrades.

each being under the responsibility of a proctor. For the latter class of students, a great club, the Harvard Union, formed a comfortable place in which to rest and lead club life. The large dining hall or Memorial Hall, furnishes board in common at favorable rates.

In many universities where the common life is less well organized than at Harvard, there are numerous student societies or clubs, which, it must be said, often have an exaggerated character of luxury or of snobbishness, but which are yet a manifestation of collective life. Certain of these societies, which they call fraternities, give themselves secret charms in their initiation ceremonies, and in their password, and are generally designated by Greek letters, standing for a mysterious motto, such as AXP, AΔΦ, ΔTΔ, ΦA, etc. The same society is represented in many universities by affiliated chapters.[1]

Another very important aspect of college life, which is connected with social life and with the training and character of the man, is the practice of physical exercise, athletics and sports. Athletics hold an enormous place, take up much time, and are in general encouraged. A vast gymnasium, often with a large swimming pool, an enormous stadium, a complete equipment for boating, are essential parts of every American university. In the psychology of the students, to belong to the football eleven, or the baseball nine, or the crews, or on a less elevated level, the tennis team, in inter-university competitions, is a title of glory very superior to the

[1] These fraternities must not be confounded with two societies of former students, spread all over America and having quite different tendencies, ΦBK (Φιλοσοφία Βίου Κυβερνήτης) which has existed for a century, admits, on leaving college, students who have been particularly brilliant in literary studies, ΣΞ (Σπουδῶν Ξύνονες) those who are dedicated to scientific research.

winning of honors of the Bachelor's degree. The boat races between Harvard and Yale on the Thames at New London, at the end of the academic year, are one of the great events of the year, like the match between Oxford and Cambridge. This is also the kind of glory which lives the longest in the memory of the alumni. A university easily finds very considerable sums to build or enlarge a stadium, to which thousands of spectators come to watch the matches between universities. Many serious minds see an undoubted abuse in the development of these matches. However, the practice of sports helps to give American youth an elegance of body and a physical vigor which one cannot but envy. It is encouraged as an effectual counter to sensual suggestions. Finally, in a general way, it accustoms them to discipline, and to team play, for later life. A mind as positive as F. W. Taylor,[1] who criticizes very severely the too great part of whim in all American education, and who opposes it to the essential condition of success in life — to do each day as well as possible what presents itself and not what pleases — considers athletics and games "one of the most useful elements of college life, and that for two reasons: 1, because they are practised with profound earnestness; 2, because they put in play not the idea of each acting at his own caprice, but of working together and of practising this coöperation in a manner like that which real life will demand." [2] "Is not the greatest

[1] *Science*, November 9, 1906.

[2] "True coöperation, coöperation upon the broadest scale, is a feature which distinguishes our present commercial and industrial development from that of one hundred years ago. Not the coöperation taught by too many of our trades-unions which are misguided, and which resembles the coöperation of a train of freight cars, but rather that of a well-organized manufacturing

problem in university life," he adds, "how to animate the students . . . with an earnest, logical purpose?"

The sociability of college life is expressed, finally, in the innumerable group or clubs which are formed. Students interested in the study of French at Harvard — and elsewhere — have formed a *Cercle français.* I was pleasantly received there. With the aid of Radcliffe college students, this *Cercle* gives each year performances of French plays, of course in French. At the Copley Theatre in Boston I saw this troupe of students play, for the benefit of our blind soldiers, *Edgard et sa bonne,* by Labiche, and a quite modern play of actual interest, *Servir,* by M. Lavedan. There is likewise a Germanic society, and very numerous debating societies, in which the young men practise public argument.

The taste for the theatre is very much alive, particularly at Harvard, where the professor of dramatic literature, Mr. G. P. Baker, instead of limiting himself to lectures *ex cathedrâ,* has his Harvard and Radcliffe students write plays which are then performed before the university families, on the little Radcliffe stage, while waiting for the generosity of alumni to make the building of a real theatre possible. Thus every year the department of dramatic literature, which is called the 47 Workshop, gives a few performances of unpublished plays, and guests are urgently asked to formulate on the spot their remarks and criticisms.

The press of a university like Harvard is another manifestation of social life. The students edit three or four papers, more or less satirical, *Harvard Crimson*

establishment, which is typified by the coöperation of the various parts of a watch, each member of which performs and is supreme in its own function, and yet is controlled by and must work harmoniously with many others."

(daily), *Harvard Lampoon*, etc. In every way, one feels the taste for collective activity.

On the whole, college life, by its relative luxury, by the spirit which predominates and by its traditions, without being aristocratic, nevertheless suits especially rich youth, who do not bring to it an ardent desire for study. The greatest individualities rarely come out of the college. They are generally self-made men. That is the case with Graham Bell, the inventor of the telephone, Edison, and most of the great captains of industry, like Carnegie or Rockefeller. Many young men go to college because their parents went there, because their families think they will form agreeable and useful relations there, finally and above all because young men are known to pass pleasant lives there. With its good qualities and its defects, the college forms a social *élite*, especially from the general point of view of culture. All factors combined, it produces the dominating personalities in most careers, and passing through college appears a weighty element to ensure success in life. This consideration of success holds a very large place in American psychology. One sees it expressed at every moment in the addresses of educators. Leland Stanford University itself depicts its goal as being " to fit young persons for success in life." Between the universities and colleges, all aspiring to grow, there is a very lively competition, and each tries to persuade the public that the sacrifices endured for the education which it gives, are a good investment for the future.

The judgment on the American college is therefore necessarily complex. It is not an institution of a purely intellectual order, nor fully answering our conception of higher education. From this point of view one may criticize it very severely, as does Mr. J. McK. Cattell:[1] "Students who complete the work of the high school at the age of eighteen can not to advantage spend the four subsequent years in a country club, where what time can be spared from athletics and social enjoyments may be given to studies that are irrelevant to their work in life. Such a system may be proper for a hereditary aristocracy of wealth, but it no longer obtains even in Great Britain, where Oxford and Cambridge are being transformed into professional universities." But on the other hand, it is not that austere abstraction, separated from life properly so-called, and built out of whole cloth by the state, without habits of active collaboration, without any spontaneous and cordial participation on the part of the public, such as are our Faculties. A throng of lively youth animates the campus and is sincerely attached to it. They carry away from it pleasant memories, oftener than the solid baggage of a scholar. Is it necessary that there should be in it an enormous number of strong scholars? Besides, the system does not at all shut out their production, any more than the development of strong scientific individualities. One may reproach it especially with being, in spite of the correctives which are applied to it, too much for the use of the wealthy classes, and with furnishing, for the average case, a culture not sufficiently deep to be fertile. That is moreover the profound cause of the college crisis of which I have already spoken, and which the examination of the other parts of the university will make more definite.

[1] *Science*, September 20, 1907.

CHAPTER VII

YOUNG WOMEN AND THE COLLEGE

Prevalence of coeducation in the western universities. Its still exceptional character in the eastern. Women's colleges. Parallelism of studies. Social results. Education and the race problem.

IN the preceding chapter it was a question of men students only of the college. But beside them we cannot ignore the women students, who are not in any degree a rare or exceptional phenomenon.

The colleges and universities contain in all, a large and rapidly increasing number of women. In 1889–90, there were 20,874 women, 38,900 in 1900–01; 77,120 in 1913–14. In the last quarter-century the number of women taking advanced studies has therefore more than trebled; and it is more than half the number of men.

In the cultivated class in moderately easy circumstances in the East, young women frequently pass through college, from the age of eighteen to twenty-two, like the young men, and take entirely parallel studies. The American woman, on this social level, has, on the whole, a more solid general culture than the man, because she takes these studies more in the true spirit of culture, and not as a means of entering as rapidly as possible into the struggle for a living. And it is women like Miss Carey Thomas,[1] president of Bryn Mawr College, who, in the evolution of the college, prove the staunchest defenders of its classical tradition, sapped by modern necessities.

[1] See notably *Congress of Arts and Sciences*, Universal Exposition, St. Louis, Vol. VIII, pp. 133 ff.

In a general way, the American woman, at the present hour, is much more emancipated from the masculine tutelage than the European woman, and that is evidently related to her education. She lives much more without help. The conditions of material life and of marriage have driven her, more than in Europe, to make sure of her own existence. She is found in a number of professions — those in which one meets her in Europe — and in others also, where among us her presence is at least exceptional. And we are astonished, on returning to France, that in our offices, in our libraries, our secretaryships to the faculties, our secondary teaching and in our administrations in general, she has not a much larger place. But the war is going to cause us to take a gigantic step in this direction. Men will have to be reserved for jobs where their strength renders them indispensable. The too numerous women will supply their places where men are not necessary.

In America, woman already has a large place as a citizen. She has all the political rights in most of the western states, and even in the Middle West; and they told me in California that, the experiment having been made, those who had been her adversaries on this ground, would make much less opposition for her today. The electoral influence of women has been beneficent, especially in municipal questions, where they have carried on a severe struggle against graft. In 1916, for the first time, a woman was elected to Congress at Washington by the state of Montana. She did not master her nerves there, it is true, as one could see at the time of the vote for the war.

To limit ourselves here to the universities and colleges, the first fact to set down is that of coeducation.

All the universities of the West practice it, and the women in the college are often as numerous [1] as the men. The prevalence of coeducation is explained by historical reasons. During the period of peopling the West, the lack of teachers, the small density of population, the great distances, imposed this organization, which is, besides, much more economical. It was quite naturally extended to higher education, when the universities and colleges were created in these regions. The oldest coeducational college is Oberlin College, in Ohio, which has used this system since its foundation in 1833.

In all of the 569 colleges and universities in the Report of the Commissioner of Education for 1913–14, 333, or more than half, are coeducational. The 236 others comprise 61 Catholic institutions, 55 for men and 6 for women; and 165 non-Catholic, 89 for men and 86 for women: of these latter only 36 are indicated as non-sectarian.

In the East, the Puritan tradition has evidently been opposed to coeducation. Yet many large institutions there are mixed. Such is the case at Cornell University (3,731 men, 463 women in 1913–14 in the college), at the University of Pennsylvania (2,226 men, 613 women), at New York University (3,019 men, 362 women), at Boston University (544 men, 613 women), and at Brown University at Providence (678 men, 203 women), and at very many other institutions (Tufts College, University of Rochester, Syracuse, Maine, etc.).

[1]

	Men	Women
University of California	2,901	1,782
" Minnesota	1,644	1,165
" Wisconsin	2,865	1,124
" Chicago	2,020	3,426
Northwestern University	573	636

Many large and old eastern universities have remained for men only, like Harvard, Yale, Columbia, Princeton, as well as Johns Hopkins, at least in the college, for Harvard and Columbia admit women graduate students.[1]

On the other hand, affiliated women's colleges, (Radcliffe, Barnard), have been organized near Harvard, Columbia, and Johns Hopkins, having the same teaching staff as these universities. This situation is not very far from coeducation, and in fact Harvard and Radcliffe students organized certain things in common, as for example the 47 Workshop which was considered above.

Among the 92 existing women's colleges, the best known are the following:

Name	Location	Founded	No. of Students	Volumes in Library	Endowment
Bryn Mawr	Bryn Mawr, Pa.	1885	467	75,000	$1,885,000
Smith	Northampton, Mass.	1875	1,549	52,000	1,790,000
Wellesley	Wellesley, Mass.	1873	1,480	80,000	2,056,000
Mt. Holyoke	So. Hadley, Mass.	1837	772	. . .	1,425,000
Vassar	Poughkeepsie, N. Y.	1865	1,077	85,000	1,520,000
Radcliffe	Cambridge, Mass.	1879	582	32,000	1,025,000
Barnard	New York City	1889	684	8,000	1,420,000
Hunter	New York City	1870	2,156	19,000	Municipal

Bryn Mawr is not only a college, but also has graduate students doing original research, and has furnished 49 doctorates, between 1898 and 1913. This college has numbered among its professors, at least in biology,

[1] Harvard does not admit women to its Medical School, contrary to Johns Hopkins. Columbia is an institution for men, but Teachers' College which merged with it, is coeducational (372 men, 1,431 women in 1913-14).

many of the most notable scientists of today: E. B. Wilson, T. H. Morgan, J. Loeb, etc. The other colleges have scarcely more than undergraduates.

The teaching in these colleges is moulded after that of the men's universities. Their studies last four years and end in a similar graduation.

From the point of view of studies, the American girl has the reputation of being more industrious than the average boy. She is much less absorbed by athletics and other pleasures. In the coeducational universities she has in consequence, very fine scholastic success, which does not altogether fail to excite masculine jealousy. It even happens, it seems, that the success and too large numbers of girls turn men away from certain subjects, especially in the literary departments.

But studies do not quite suffice to give a picture of college life for women. In the exclusively women's colleges this life naturally has the most character.[1] I had a chance to visit only Wellesley College, near Boston. Its equipment is magnificent. The college is an immense park, of more than 400 acres, in a charming site, with a beautiful lake bordered by wooded hills. Over this great domain the college buildings are scattered laboratories, an observatory, a chapel, a building for the fine arts, another for music, professors' houses, and beautiful buildings where the young women live. The main building, which has just been rebuilt after a fire, is truly sumptuous. We have no idea in France of establishments of such luxuriousness. If we add that, in these colleges, the young women enjoy a very large liberty, that they have organized a common life anal-

---

[1] In the mixed universities the young women of course have their special dormitories or residential colleges.

ogous to that of the men, sometimes with clubs and
sororities (parallel to the fraternities), one conceives
without difficulty that the years in college are for them
also a "good time," and that they have no apprehen-
sions in devoting to it four years of their youth.

But we must ask also what are the results from the
social point of view. A not negligible proportion of
the young women who study in the colleges are destined for teaching. This proportion is even strong in
some colleges, as at Bryn Mawr. But the majority of
the 77,000 women students are scattered among various careers, or simply wait for marriage. Passing
through college has largely emancipated the American
woman. She is cultivated; she has a free mind; she
interests herself willingly in very varied things; in
particular in problems of public usefulness, often in a
somewhat fearsome manner. Yet one cannot help
thinking that the life which they have led,[1] during the
period of their studies, has tended to develop in them
luxurious tastes which in many cases oppose a serious
obstacle to family life.

Besides, that is a question for all American society,
and especially that of the East, to solve. The birth-
rate there is very small, much less than in France, on
which subject the Americans are often alarmed, without always taking account that their case is much worse,
but that immigration has up till now furnished them
wherewith to fill the vacancies. In Massachusetts from
1877 to 1891, the recent immigrants have furnished an
excess of births over deaths equal to 526,987 individ-

---

[1] The minimum expenses of a Wellesley student are about equivalent to
those of a Harvard student.

uals, while the native population showed, during the same period, an excess of deaths over births equal to 269,918. The American population of ancient source, depositary of English civilization and Puritan tradition, is thus threatened with rapid disappearance. This sterility is evidently voluntary, at least in general, and the general comfort of living and the economic exigencies which that entails, are one of the principal causes. These causes apply to very varied classes of the population. But the statistics show that the problem is very grave, as it concerns the colleges. Students of population and educators are keenly disturbed over it. College education tends to aggravate the evil rather than to remedy it.

Mr. R. S. Sprague,[1] professor of social economics at the Massachusetts College of Agriculture at Amherst, from whom I have borrowed the figures below, denounces at once the high school and the college. "The

Date of Graduation	Per cent Unmarried in 1916	Per cent Married	Average Number of Children per Married Graduate	Average Number of Children for the Total Number of Graduates
1842–49.........	14.6	85.4	2.77	2.37
1850–59.........	24.5	75.5	3.38	2.55
1860–69.........	39.1	60.9	2.64	1.60
1870–79.........	40.6	59.4	2.75	1.63
1880–89.........	42.4	57.6	2.54	1.46
1890–92 [2]......	50.0	50.0	1.91	0.95

[1] *Journal of Heredity,* Vol. VI, 1915, p. 159.

[2] The statistics are ended at 1892, in order to take in only women whose period of maternity may be considered as ended.

high schools," he says, "have turned their backs on the family. They prepared our children for the college, for the parlor, club, and travel, but did not recognize the demands of the workshop, kitchen, and nursery where the greater part of the average parent's time and energy must be spent." "College education," he says again, "gives independent women, but women whose individual superiority is acquired at the expense of the race. The statistics relative to the number of children of former students of the women's colleges are very lamentable. For example, there are those of Mount Holyoke College, which is the oldest.

Vassar College furnishes, between 1867 and 1892, 959 graduates, of whom 509 have married, or 53 per cent, and have had 973 children, or about 1 per graduate. Of these 959 women, 451, or 45 per cent, have taught; among the latter 166 have married, and have had 287 children, or 1.73 per capita; moreover, 343 graduates, who have not taught and who have married, had 686 children, or 2 on the average.

Bryn Mawr, which is more recent, being founded in 1885, cannot furnish comparable statistics for the classes 1888 to 1900, 165 graduates, 45 per cent, were married on the first of January, 1913, and had had 138 children, or 0.84 per married graduate, and 0.37 per capita, of the total of graduates. These figures may yet be improved, but are deplorably small.

Wellesley, for its classes from 1879 to 1888, furnishes 55 per cent of marriages among its graduates and 60 per cent of the total of its students, whether graduates or not. These marriages have furnished an average of 0.86 children per graduate and 0.97 for the total of the classes (1.56 per married graduate, and 1.62 for the total of students married). The statistics for the honor-girls give still smaller figures.[1]

The principle of women's colleges has sometimes been given as a cause, and the preceding low figures attributed, at least in part, to a mentality which the absence of coeducation would develop. That would plead for the general adoption of the latter, which, besides, is more economical. But some statistics quite recently published,[2] give for the women coming from Syracuse University,[3] figures absolutely agreeing with those for Wellesley College.

	Wellesley	Syracuse
Average number of children for the total number of graduates.	0.86	0.88
Average number of children for each married graduate.	1.56	1.60

Therefore coeducation or its absence does not seem to be an important factor in the question. Accordingly, one can hardly refuse to admit the intellectual development of the American woman and her taste for culture,

[1] The *Journal of Heredity*, from which these statistics are taken, has calculated for thirty years, 1861–90, the average fecundity of graduates of Yale and Harvard. The percentage of marriages oscillates between 75 and 80 per cent, even for recent years. The average number of children per family has gone down from 2.24 for the first decade to 1.87 for the last. The average number of children, calculated on the total of the graduates, is 1.54.

[2] *Journal of Heredity*, May, 1917, H. J. Baeker, Coeducation and Eugenics.

[3] By comparing statistics of men and women students of Syracuse University, conditions of the environment being similar, we find:

	Men	Women
Per cent of married graduates.	81.0	57.0
Average age of marriage.	28.8	27.7
Per cent of marriages between students of the university.	3.46	2.06
Per cent of sterile marriages	20.0	28.0
Per cent of marriages with two children	45.0	42.0
Average number of children to each family	2.06	1.46
Average number of children to each graduate.	1.66	0.83

but she has been led into a very strong individualism and turned away from the prosaic realities of life. The general wealth and perfection of material life have also had in their own way a considerable effect.   Mr. Sprague says energetically, but with reason, "Women are the capital of the race. The farmer who employs his land as a golf course or a hunting preserve, instead of cultivating it, is surely going to ruin. Likewise a civilization which employs its women as stenographers, clerks and professors, instead of making mothers of them, is running to the ruin of the race."

In what specially concerns the marriage of women college graduates, the law of supply and demand seems to work against them.[1] "They are not prepared psychologically or technically for the occupations of family life, and seek these only under special conditions, which drive many men away from them." [2]

The movement which leads women toward a high intellectual culture is not evil in itself, but it ought to have important correctives, in the teaching given, so as to restore to its true place, which should be the first, the rôle of woman as wife and mother.[3] Superior in-

---

[1] Mr. McK. Cattell gives the following statistics: American men of science who have married women graduates have had on the average 2.02 children; those who have married women who have gone to college, but not graduates, have had 2.12, and those whose wives have not been to college have had 2.32. Many intelligent men prefer young women whose education has turned them toward the realities of life; and that is one of the chief causes of the lowering of the percentage of marriages among women college graduates.

[2] Sprague, *loc. cit.*

[3] According to R. K. Johnson and R. Stutzman, *Journal of Heredity*, 1915, the women's colleges offer an obstinate resistance to introducing into their programs domestic education, and especially anything which concerns the care to be given to childhood.

dividuals only — and they are naturally very rare — can pretend that they are exempt from this law, and the highest virtue of a woman, even an educated one, is still to make sure the future of the race. That is what Napoleon I curtly replied to a question by Mme. de Staël, in the name of good sense, which never loses its rights.

CHAPTER VIII

GRADUATE SCHOOL OF ARTS AND SCIENCES

Relations with the college. Development. Degrees. Master of Arts. Doctor of Philosophy. The doctorate in the principal universities.

THE Graduate School of Arts and Sciences is, as we have seen in Chapter II, a sort of prolongation, which has been grafted upon the college, and which is the true equivalent of our Faculties of Letters and of Sciences.[1]

In general it has not a very well-marked physical individuality. Its staff, that is to say its faculty, is composed of the same men as that of the college. It has only an administrative individuality, indicated by a special dean, who manages its business. Its departments are those of the college, and as we have seen, are not absolutely distinct, at least at Harvard, and at many universities. Certain courses are for graduates and undergraduates, and others primarily for graduates. In certain universities, however, the distinction is better marked. That is the case at the University of Pennsylvania, and especially at Johns Hopkins, which is properly speaking exclusively a graduate school, the college which is its satellite being quite distinct. At Bryn Mawr the arrangements are parallel to those at Johns Hopkins.

---

[1] The two groups of studies are in general united, as the term Arts and Sciences moreover indicates. Sometimes, however, there are two distinct organizations. That is the case at the University of Chicago, which has a Graduate School of Arts and Literature, on the one hand, and on the other hand, the Ogden School of Sciences.

Clark University at Worcester is also a pure graduate school, which now has a college connected with it.

By virtue of this organization, the graduate school is not in general of a radically different character from that of the college. The courses primarily for graduates are only more specialized and presuppose previously acquired knowledge. Each department has a smaller or larger number of them, and the flexibility of the organization in departments permits them easily to arrange new courses on the parts of science which have just developed. We can give a clear example of it for biology. Since 1900 a special branch of this science has been constituted, which has its origin in the already ancient works of Charles Naudin and Gregor Mendel, the experimental study of heredity, by way of crossing neighboring varieties. This study, whose centre is heredity, called Mendelism, has formed the object of numerous researches in the last fifteen years, and constitutes today what is called genetics. Genetics is abundantly taught in the large American universities, at once in an elementary form for undergraduates, and in a more advanced way for graduates. At Harvard this instruction is given, on the zoölogical side, by Professor W. E. Castle, on the botanical side by Professor E. M. East. It is likewise represented in all the great universities. These last years, genetics was taught in 51 important establishments, representing 3000 auditors of these courses; in 15 of them there were 140 graduates engaged on original researches in this domain.

While finishing their instruction in the various parts of a science or group of sciences, the graduate students are under the more special direction of a professor, who guides them in their researches, and the graduate school is before all, the collection of laboratories and seminars corresponding to the various sciences. It is therefore above all a particular aspect of the general activity of the college professors.

It is a recent differentiation, which goes back scarcely over fifty years, even in the universities in which it is the oldest. The first trace of organization of studies for graduates at Yale is proven to be in 1847, and the graduate school was definitely established there in 1872. At Harvard from 1860 there were graduates about Louis Agassiz. The doctorate was instituted there in 1872. There were 28 graduates in 1872, 111 in 1889–90.[1] There are at present 500 of them.

The graduate school is well represented in only a small number of universities, and one may take for one of his criteria, the number of the students who are enrolled in them (in 1913–14). I have set forth the total number of the students of the university, so that one can get an idea of the relative importance of the graduate school in the whole. These figures, however, give the total of graduate students, and for certain universities, such as Harvard and Chicago, they include organizations distinct from the Graduate School of Arts and Sciences (for example, at Harvard, the Graduate School of Business Administration, or advanced school of commercial studies). (See table on opposite page.) We find, besides, 25 institutions which show, from their statistics, graduate students varying in number from 200 to 80. Moreover, in 1913–14, the total of graduates given in the Report of the Commissioner of Education, was 12,871, of whom 8656 were men and 4215 women.

[1] E. D. Perry, *Monograph on Education in the United States*, Vol. II, p. 6, Exposition, Paris, 1900.

They represent, therefore, numerically almost one-twentieth of the total number of American students.

Universities	Total No. Students	Total of Graduate Students	Men	Women
Chicago.....................	7765	1799	1081	718
Columbia..................	5112	1568	942	626
Columbia Teachers' College....	1803	681	362	319
Harvard...................	4912	795	713	82
California.................	6028	707	404	303
Wisconsin.................	4719	480	343	167
Cornell....................	5015	383	327	56
Pennsylvania...............	4686	337	308	129
Michigan..................	5520	298	230	68
Illinois....................	5094	285	242	43
Yale......................	3189	258	202	56
Johns Hopkins.............	852	229	189	40
Princeton.................	1600	176	176	....
Leland Stanford Jr..........	1906	224	168	56
Minnesota................	4958	166	117	49

The work of the graduates is rewarded by two degrees, that of Master and that of Doctor.

The title Master of Arts, M.A., or Master of Science, M.Sc., can be obtained by one year's residence at the University, after graduation, during which the candidate works on a program traced out in advance and approved, at the time of enrolling, subject to certain rules as to extent and variety. The rules vary from one university to another. There is not always an examination. At the University of Chicago, the candidate must write a little dissertation, which is deposited in the archives. At Cornell University, they require both an examination and a small original work. To sum it up, the studies for the degree of master require a certain specialization, with apprenticeship in original work. We can consider that a master from a good university corresponds approximately to one of our *licenciés ès sciences*, holding a diploma for advanced studies. In 1913–14, about 4700 Master's degrees were conferred.

The Doctor's degree, Doctor of Philosophy, Ph.D., or Doctor of Science, Sc.D., is generally acquired by three years of studies as a graduate, the year for the Master's degree being included in these three years. The candidate submits a subject for study to a committee, and must write an original thesis. The trials are composed specially of an examination on two accessory or minor subjects, and on one principal or major subject. The thesis is generally printed, though this is not absolutely obligatory. The doctorate has been more or less modeled after the German doctorate; but here it is not the normal termination of the studies of all the students. It is rarely won with less than seven years' residence at the university, at the age of twenty-seven or twenty-eight. It seems to me to represent, on the whole, an effort rather comparable to that which our doctorate in law exacts, the value of the comparison depending on the measure in which the two kinds of studies can be compared.

This comparison with our doctorate is, moreover, an important and present one. One of the reasons — it is not the only one — why the Americans have flocked to the German universities so much, is that they can there become a *Doctor Philosophiae*, and bring back this degree to America as a certificate for their work.

No barrier, like our licentiate, drives them away. It often happened that they even passed off in Germany theses written in America.

Today, when the war is bringing forth a great current of sympathy from America toward France, the Americans would like us to accord them the same facility. Our state doctorate is practically inaccessible, and foreigners understand imperfectly the distinction between the two doctorates, one state, the other university. In my opinion, there would be ground, while maintaining the state doctorate as it is, for our nationals, to suppress the antecedent condition of the licentiate for foreigners, and to accept what would be the equivalent American degree.[1]

University	Number of Doctorates in 1914–15	In Science	Annual Average, 1898–1907	Total Number 1898–1915	In Sciences
Columbia..........	70	27	32.2	835	329
Chicago...........	79	53	35.6	780	414
Harvard...........	58	33	33.8	709	296
Yale..............	36	20	31.8	590	267
Johns Hopkins......	31	23	30.5	536	324
Pennsylvania.......	34	11	22.5	458	177
Cornell...........	31	26	18.1	452	317
Wisconsin.........	21	8	8.6	258	108
Clark.............	12	10	8.7	180	162
Michigan..........	26	15	6.9	158	76
Illinois...........	23	17	5.0	122	81
California.........	22	16	3.3	120	89
Princeton	12	4	2.6	111	49
Stanford..........	5	2	1.4	47	32

[1] One of our American colleagues, who was a student in the *École Normale Supérieure*, a little more than twenty years ago, and who has kept a very good memory of it, quite recently wrote me these lines, which it seems to me interesting to quote here: "In particular it would be very desirable that a committee in France should examine the catalogues (of the principal American Universities) and should learn the conditions exacted for the doctorate in America, *so that the Sorbonne might offer to American students a doctorate*

These are the figures relating to the universities which are most important for the sciences.[1]

The statistics show that about fifty universities or colleges give the Doctor's degree, thirty or thirty-five of them in a regular manner.[2] In 1914–15, 556 doctorates were granted, which is a number double the average of the decade 1898–1907. If we wish to consider the sciences separately, there are other tables.[3]

Of course we must not exaggerate the value of figures in matters such as that which occupies us. The value of scientific work is more important than their number. The number of doctorates given is one of the tangible elements on which the universities count for establishing their reputation with the public; driven by the spirit of mutual rivalry, they may tend to pad this number by an excessive indulgence.

having about the same degree of difficulty as our own. I think that the *Doctorat d'Université* answers this requirement. But unhappily very few Americans go to the Sorbonne. As a former student of the *École Normale*, I wish this school could establish a section for foreigners. Besides, I think it would be very desirable to organize a centre of information for our students. *The chief difficulty in coming to study in France is that at the Sorbonne there is no suitable machinery to attend to the American students,* or in any case, if this machinery exists, it is unknown in America. Our students *feel more or less abandoned when they come to Paris. If they go to Göttingen or to Berlin, things are organized in such a way that they easily find their place in the institution and get what they want."*

The same requirements are expressed identically in other letters which have been sent to me.

[1] New York University, Boston University, and Bryn Mawr, furnish a very considerable number of doctors, for the whole of the period, but have only a very small importance as regards the sciences.

[2] At Cornell in 1910, there were the following numbers of candidates for the doctorate: 53 chemists, 27 botanists, 24 physicists, 19 zoölogists, 20 geologists.

[3] The tables are given in *Science*, October 22, 1915, pp. 555 ff.

But the figures become more significant when compared with a past of twenty years. We see how rapid

	Average 1898–1907	1914–15	Total 1898–1915
Chemistry...................	32.3	85	838
Physics.....................	15.5	31	366
Zoölogy....................	15.2	32	348
Botany.....................	12.6	40	315
Psychology.................	13.5	22	309
Mathematics................	12.1	23	297
Biology....................	7.1	26	190
Physiology.................	4.1	8	97
Agriculture................	1.0	9	71
Astronomy..................	3.4	7	71
Bacteriology...............	1.4	4	44
Anthropology...............	1.0	6	33
Anatomy...................	0.9	5	27
Paleontology...............	1.6	2	25
Engineering................	0.8	2	19
Pathology..................	0.5	2	19
Mineralogy.................	0.6	1	11
Geography.................	0.1	3	8
Metallurgy.................	0.3	1	6
Meteorology................	0.1	0	1

is the development of the universities, and how it is made in all directions. In particular, the state universities are rising from the utilitarian level on which they began, and tend to take an honorable place beside the great private universities of the East.

The students who are preparing for the doctorate are, in a large majority, future college or university professors. That comes out with extreme clearness in the catalogue of Harvard theses from 1873 to 1916. Almost all of these doctors are now teaching or have taught in universities or colleges, or are connected with museums or with governmental scientific services. Only those young men who wish a scientific or pedagogical career push on to the doctorate.

The social construction of the graduate school is therefore quite different from that of the college, as is naturally to be expected, and the material importance of the rôle which the college has in the life of the university is understood.

We will return to the conditions of scientific research at the opening of the second part.

CHAPTER IX

THE PROFESSIONAL SCHOOLS

First group: Theology, Law, Medicine, Dentistry, Pharmacy. Second group:
Schools of Pedagogy, Teacher's College at Columbia, the School of Education
at Chicago. Schools of Fine Arts. Architecture. Schools of Journalism.

THE parts of the university which are still to be considered are called by the Americans professional
schools, because the teaching in them is preparatory to
a definite career. I will touch but lightly upon those
which correspond to our Faculties of Law and of Medicine, because I do not feel that I am a sufficient authority, and because they are less interesting to us than
the others. A large number of professional schools have
been created in an independent way, and at their
origin were conceived with a strictly utilitarian end, in
order to train professional men as rapidly as possible.
Gradually, they are merging into the universities.

For medicine and law in particular, it is only quite
recently and under the impulsion of universities like
Harvard, Johns Hopkins and some others, that they
have endeavored to get students having considerable
preliminary knowledge, by requiring a period of study
in college before entrance to these schools.

FIRST GROUP. SCHOOLS OF THEOLOGY, LAW, MEDICINE, AND THE ASSOCIATED PROFESSIONS

First the statistics which show the number of schools
of these various specialties, as well as the size and

quality of their faculties and of their student bodies in 1913–14.[1]

Faculties	Number of Institutions	Number of Professors	Total No. of Students	Graduates	Per Cent of Graduates	Women Students
Theology......	176	1516	11,269	417	3.7	580
Law..........	122	1471	20,958	4215	20.1	522
Medicine......	100	6955	16,920	2468	14.5	835
Pharmacy.....	72	744	5,930	91	1.5	280
Dental Schools	50	1532	9,315	199	2.1	185

1. *Schools of Theology*

For theology, I limit myself to showing the figures above.

The state universities have no faculty of theology. As for the private universities, one might distinguish their faculties of theology by noting whether the institution is a church school or undenominational.

2. *Law Schools*

The course of studies in these schools is generally three years and leads to a *baccalaureate* (Bachelor of Laws) given, following examinations passed every year. There is also a degree of doctor; at Harvard it is obtained by a fourth year of study and special examinations.

[1] By way of comparison, the statistics for 1898.

Faculties	Number of Institutions	Number of Professors	Total Number of Students	Women Students
Theology....................	165	1070	8317	198
Law.......................	86	970	11,783	147
Medicine....................	156	5735	24,043	1397
Pharmacy...................	52	412	3525	174
Dental Schools..............	56	1513	7221	162

Harvard is the only university which requires all its law students to be previously college graduates. The University of Chicago requires either the Bachelor's degree or three years of effective and satisfactory work in a college. It is the same at Columbia. There, however, they admit a less precise equivalent, in the form of a certificate of satisfactory previous study and instruction, in an advanced American or foreign institution.

Here follow some figures relating to 1913–14, which show the proportion of graduates among the law students in various universities.

You will observe that none of the large universities has a very numerous school of law, as one would have expected.

University	Total Law Students	Number of Graduates	University	Total Law Students	Number of Graduates
Harvard	694	691	Stanford	158	20
Chicago	313	255	California	223	99
Columbia	493	410	Wisconsin	220	50
Cornell	293	12	Illinois	121	9
Pennsylvania	374	155	Michigan	612	79
Yale	133	103	Minnesota	176	17

Law students, at least at Harvard, are known to be extremely hard working; the spirit is quite different from that of the college. I am not competent to speak of the instruction. From what I have been told, its character is less doctrinal than practical.

A certain number of schools, of which some are very large (that of Georgetown University at Washington has a thousand students), have only evening courses.

3. *Schools of Medicine*

The teaching of medicine in the United States, until a very recent period, was very unsatisfactory, as Americans admit. Physicians lacked general scientific education. "In medicine as in politics," writes Professor Howell of Johns Hopkins, "a country finds, on the whole, the kind of services which the majority considers necessary, and progressive men have much difficulty in leading this majority to change its ideas."

The first schools were those of Philadelphia (today, University of Pennsylvania), and of King's College (Columbia), at New York, which go back to the eighteenth century. The Harvard Medical School was founded in 1782.

About thirty schools were organized in the first half of the nineteenth century, and a hundred in the second half. The great defect of these schools — and it is not entirely overcome today — was in not exacting any previous general instruction from their students. Many schools competed in reducing the difficulties of the studies. The very great liberty granted by law to the medical profession in the absence of a state examination, did not tend to remedy these defects in the training of the physician. On the other hand, numerous extra-medical systems, real sects, pretending to cure, have been very much in favor with the public, and even today, Christian Science has a great influence even in the most cultivated parts of the United States, such as Massachusetts.

Of the 156 schools of medicine existing in 1900, 82 were still independent of universities. These 156 schools are today reduced to 100, of which 10 are homeopathic and 4 eclectic. In certain large cities there are several independent medical schools, either attached to universities or autonomous. Chicago has no less than 7, New York (with Brooklyn) 6, Philadelphia, 5, Washington, Baltimore, Boston and St. Louis, 3. None of these schools is very large. The largest have 600 to 700 students. Many have only a hundred, and some are rudimentary. Oakland College of Medicine, near San Francisco, had, in 1913–14, 16 students and 44 professors, or instructors. Several of the most important universities have no medical school. That is the case with Princeton. Yale has only about 50 students in its department of medicine, which, however, numbers 57 professors, or instructors. At the University of Chicago, there were in 1915–16, 599 students in medicine (68 of them women), but the university itself has no real faculty of medicine. It has organized only the scientific part of the medical course (physics, chemistry, physiology, anatomy, pathology), but possesses no clinical instruction or hospitals. It has made an agreement with the Rush Medical College, which is, in fact, its school of medicine, but with which it has no official connection. The scientific courses of this college have been transferred to the University of Chicago, and the students of the latter take their strictly medical studies in the former.

The progress realized in the quality of medical studies has been due above all to the influence of Johns Hopkins and Harvard Universities. Johns Hopkins admits only college graduates to its medical school proper. Cornell and Yale require three years of college. Harvard admits graduates, or students having completed at least two years in college, under conditions satisfactory from the scientific point of view. The rule which is becoming established is to require two years of college, before the strictly medical studies, which themselves last four years. Of the latter, the first two are of a more scientific order, and thus form, with the two years of college a whole similar to the classical Bachelor's degree.

The following figures [1] show the proportion of graduates among the medical students in the great universities.

University	Total	College Graduates	University	Total	College Graduates
Johns Hopkins.	360	360	California....	124	25
Harvard......	308	290	Illinois.......	351	26
Chicago.......	462	87	Michigan.....	288	94
Pennsylvania..	284	156	Wisconsin....	83	11
Columbia.....	355	252			

The Harvard Medical School, rebuilt in 1907, at Boston, in the neighborhood of the principal hospitals, is magnificently equipped. The medical schools of the large universities established in the country or in small towns, are generally separated from the rest of the university, in order to be placed in a large city. Thus the schools of medicine of the universities of Illinois (Urbana), and Northwestern (Evanston), are in Chicago, and that of Cornell University (Ithaca), is in New York City.

[1] Women are still excluded from many medical schools (Harvard, Columbia, Yale, Pennsylvania). They are admitted to that of Johns Hopkins, because of a special condition in a legacy. There are a certain number of special schools for them. At present I find only one mentioned, at Philadelphia (Woman's Medical College of Pennsylvania), with about 100 students.

The duration of the medical studies is at present four years. This, for example, is the arrangement of subjects at Harvard:

First year: Anatomy, Histology, Embryology, Physiology, Biological Chemistry.

Second year: Bacteriology, Pathology, Prophylaxis and Hygiene, Pharmacology, Medicine, Surgery, Neurology, Dermatology.

Third year: Medicine, Surgery, Obstetrics, X-rays, Syphilis, Psychiatry, Legal Medicine.

Fourth year: Numerous scientific or clinical specialties, at the choice of the candidates, and four months as clinical clerk in a hospital.

The degree granted at the end of the studies is that of *Doctor of Medicine*. Harvard also gives that of *Doctor of Public Health* to doctors of medicine who work an additional year on a special subject and write a thesis embodying original researches.[1] Finally, it has superimposed upon its medical school, a *Graduate School of Medicine*, whose work consists of courses and especially of original scientific research in the laboratories. It has constituted, within it, a special school of tropical medicine, which Professor Strong directs.

To sum up, the teaching of medicine is still in a period of great transformation, and throughout the country is very heterogeneous. Mr. Howell, of Johns Hopkins, thinks that the three chief problems actually under discussion are the opportuneness of a third year of clinical studies, the better adaptation of clinical professors to the scientific method, and the means of getting them to devote themselves more to their teaching.

---

[1] Besides, studies on branches of the medical sciences may be combined with college studies and lead to the degree of A.M. or Ph.D.

4. *Dental Schools*

While the United States remained very backward in medical studies, the practice of dentistry — in which the mechanical has a very large part — developed to a very high degree there. Numerous schools were organized, a large number of which are attached to universities. There are at present 50, with 9000 students (more than half the number of medical students). The dental school is a constituent and sometimes an important part of most of the large universities. That of Northwestern University numbers, in 1913–14, 615 students (46 professors); that of the University of Pennsylvania, has 558 students (63 professors); that of Harvard has 193 students (59 professors).

These schools are very well equipped. The length of their studies is three years, and beginning with 1917–18, will be extended at Harvard and in a certain number of other schools which have formed an association to four years. At the Harvard dental school the studies of the first year, anatomy, histology, physiology, chemistry and dental physiology, and general pathology, are all taken at the medical school. The degree granted is that of *Doctor of Dental Medicine*.

5. *Schools of Pharmacy*

These schools, at present numbering seventy-two, are for the most part, attached to universities, of which they form a special college. This college, however, is lacking in a certain number of large private universities (Harvard, Chicago, Johns Hopkins, Pennsylvania), without doubt because the schools of pharmacy existing in their respective cities have kept themselves independent. In

New York, the College of Pharmacy, formerly distinct, has merged with Columbia.

The instruction in this last named school of pharmacy is given in two degrees. In two years of study it trains ordinary pharmacists and chemist-pharmacists of the university. A third year, open to the graduates from the first course, leads to the degree of *Doctor of Pharmacy*, and gives the technical instruction necessary for the new work of pharmaceutical laboratories (microscopic, biological analysis, etc.).

SECOND GROUP. PEDAGOGY, ARCHITECTURE AND FINE ARTS, JOURNALISM

6. *Schools of Pedagogy*

The theory and technique of education are a subject of study very much in favor in the American universities. Almost all have at least a department of education in the college, in which the most diverse questions of pedagogy and of teaching are methodically studied. At Harvard twenty-four courses (for a period of two years) are offered in this division, on the general problems of education, psychology, history of pedagogy, theoretical and practical organization of teaching, in its various stages. Certain of these courses include visits to schools or even practice in teaching there. Finally, the attention of the students is drawn to a list of courses which are given in other parts of the university, such as American institutions, sociology, philosophy and psychology.

In a certain number of universities, there is a real school of pedagogy, more or less independent. The largest is that of Columbia, at New York, which bears the name of Teachers' College and which, in 1915–16, had 1972 students.[1] Teachers' College is an institution which was founded independently in 1888, and which merged with Columbia in 1898, while retaining a very broad autonomy. This college has its special board of trustees. It is, moreover, very heterogeneous. In fact it includes two distinct schools.

One is a school of pedagogy, in the strict sense, for the thorough study of psychology, sociology, the history and philosophy of education, administration of schools, and the various types of instruction (secondary, technical, elementary, kindergarten). It is the professional institute of pedagogy at Columbia, quite as much as the school of mines or of medicine is professional. The instruction in it is given in two stages: 1, two years of pedagogical instruction, to which students already having had two years in college are admitted, and it grants the degree of *Bachelor of Science in Education*.[2] 2, more advanced instruction, a prolongation of the preceding, constituting, in brief, a graduate school, and leading to the degrees of *Master* and *Doctor in Teaching*. This advanced instruction brought together, in the last few years, about 350 graduates. In 1911–12 672 students of Teachers' College have been appointed to various teaching positions, 110 of them in colleges or universities.

The other school comprised in Teachers' College bears the name of *School of Practical Arts*, and has quite a different character. It is also open to men and women with two years of college, and its aim is to realize a

[1] The women are in a very large majority; of 1803 students in 1913–14, there were 1431 women and 372 men.

[2] It is clear that the plan of this instruction is modeled on that of the general form of the college, as regards the duration of the studies and the degree awarded.

mixed type of higher education, uniting liberal culture and technical instruction in very varied directions, according to choice; industrial arts, domestic arts (feeding, cooking, dressmaking, domestic chemistry, physiological chemistry, nutrition, nursing, hygiene, fine arts, music, physical education, etc.). The school has laboratories and studios permitting practical work. Sixty-four courses were given in 1911–12. At the same time, it prepares teachers for these various branches.

At the University of Chicago, the *School of Education*, which had, in 1915–16, 1394 students (1196 of them women), has, in brief, a plan rather like that which we have just seen at Columbia. It includes, in fact, four divisions: 1, an advanced section for graduates; 2, a *College of Education* — a professional school for the training of teachers of secondary and primary branches — parallel to the classical college, conceived on the same plan, but with specialization of studies in pedagogy;[1] 3, the *University High School;* 4, the *University Elementary School.* These last two play, for secondary and primary education, the rôle of the schools annexed to our primary normal schools; the students of the first two sections are trained practically in teaching. The students of the School of Education, on the other hand, may take any of the University courses.

The two examples of Columbia and Chicago show how broadly pedagogical problems are conceived. These two schools are the most important; but some exist in other universities, which have 300 to 500 students.

In the training of teachers, a very debatable but very

[1] There are special sections for the manual arts and for domestic economy, as at Teachers' College.

interesting tendency will be noted, not to establish water-tight bulkheads between the three orders, primary, secondary, and higher instruction. The results ought to be studied, and I have no data in this regard.

7. Schools of Fine Arts, Architecture, Music

Architecture is taught in a goodly number of universities, in which it constitutes a special school. In the universities of Pennsylvania, Illinois, and Cornell, this school includes as many as 250 students. It is frequently subdivided into two, one for architecture proper, and the other called *School of Landscape Architecture*. This last specialization has an important rôle in America, by reason of the creation and development of cities. In the West especially, the traveler is struck by the uniformity of the plan of the new cities, which rests on fixed principles: the streets are laid out in a broad fashion, and hygiene is seriously studied.

At Columbia the School of Fine Arts is subdivided into three parts, architecture, music, and design. This last is scarcely more than a project. At Harvard the Fine Arts courses (to the number of about forty), or of music [1] (twelve), form a department in the college. The school of architecture alone is distinct.

The University of Pennsylvania, and Yale, and also universities of Illinois and Wisconsin, have each a special music school.

These very summary indications suffice to show the extended range of the universities in these directions.

---

[1] Notably, there is a course on Vincent d'Indy, Fauré and Debussy.

8. Schools of Journalism

This type of school, of a quite modern character, was first brought into being at Columbia (136 students in 1913–14), thanks to a gift of one million dollars, made to the university by the owner of the *World*. Later, its example has been followed by some other universities (Wisconsin, Indiana, Missouri, Tulane, New York University).

The aim is to train young men for the journalistic profession, and also to perfect journalists already at work, in the practice of their profession. The course of studies, for the former, is four years, and leads to a degree of *Bachelor of Literature*. The program includes the following subjects: English, German or French, European literatures, history, philosophy, economic sciences, history and principles of science, and technical courses (reporting, interviewing, publishing, etc.).

CHAPTER X

THE PROFESSIONAL SCHOOLS

Third group: Advanced Schools of Commerce. Harvard Graduate School of Business Administration. Chicago. Philadelphia. Engineering Schools: Origin, the Morrill Act, and the Colleges of Agriculture and Mechanics. Independent Schools of Technology. The various engineering specializations. Practical character of the instruction. Schools of Agriculture: Rôle of the Morrill Act. Colleges of Agriculture. Cornell, California, Illinois Universities, etc. The Agricultural and Mechanical Colleges. Veterinary Schools.

I CONSIDER together these three kinds of schools, which tend to assume a large and more and more individualized place in many American universities. These are the ones which are most foreign to our notion of a university. Their existence and their rapid growth are in direct relation to the character and needs of American society, and to the fact that the universities remain in close contact with the general life of the nation.

9. *Schools of Commerce*

Commercial schools of a lower order, training business employees or clerks, are extremely numerous in the United States. The Report of the Commissioner of Education shows, in 1914, 3618 schools where one may thus prepare for business, with 346,770 students. The universities, faithful to their general aim, have proposed to train the general staff of this commercial army, the leaders in this domain as in the others. They have entered in different degrees on this path; those which have advanced the farthest are those in the great cities of the East.

The University of Pennsylvania at Philadelphia, received, in 1881, a gift of $100,000 to develop advanced commercial instruction, and organized with this in view, a school named, in honor of the donor, *Wharton School of Finance and Economy*. It had 1889 students in 1915–16. The University of Chicago, from its first years, has had a school of *Commerce and Administration*, which has at present 200 students. New York University has a School of Commerce attended, in 1915–16, by 2639 men and women. I shall indicate also the figures relative to this category of students in the following universities: Pittsburgh (916), Northwestern (741), Wisconsin (542), Illinois (527), California (308).

Even Harvard, the most classical of the universities, organized a school of this kind in 1908, but has tried to make a superior type out of it. It has required a Bachelor's degree for entrance and has made of it the *Graduate School of Business Administration*, which had 182 students in 1915–16. The courses in it last two years and lead to a master's diploma. The students who enter this school at Harvard have already specialized with this intention, during their two last years in college, by choosing studies relating to economic questions. The instruction in the school includes courses on accounting, commercial law, marketing, factory organization (in particular there are courses on the Taylor system), general commercial practice, exporting, banking and finance, insurance transportation (administration and development of railroads), printing and publishing, public works, lumbering. It gives technical knowledge concerning the various branches of business to young men already trained by the general culture of the college.

At Chicago, the School of Commerce and Administration is an undergraduate school, parallel to the classical college, and extended by an advanced section for graduates. During the four years, the instruction includes fundamental required courses and elective courses, which the students choose according to their intended careers. The school is moreover divided into four sections: business, commercial teaching (a section training professors of elementary business schools), secretarial course, social service. In this last section there are numerous courses on the various social problems (public and industrial hygiene, economic legislation, municipal legislation, criminality, prostitution, immigration, study of the various ethnical types, sociology, trade unionism, games, etc.). At the same time the students may take advantage of the courses in the university. This is an organization of great flexibility, providing preparation for interesting social activities.

At Philadelphia the Wharton School is also an undergraduate college, parallel to the classical college, requiring the same conditions for entrance and offering a four years' course. The program is extremely broad, and has for its reward the diploma of *Bachelor of Science in Economics*.

I limit myself to these examples. They suffice to show the end, which is to give future business men a culture comparable in breadth to the classical culture, but adapted to their needs. It will be noted that the general mold into which these various practical adaptations are run, clings to the forms of the old college.

10. *Engineering Schools*

The engineering schools, are, at the present time, one of the essential and characteristic elements of the American universities, and are among those which are every day taking on a wider scope. That there should be great engineering schools in the United States is not surprising; but the interesting fact is that the universities should have understood the utility for themselves of keeping this branch which is so important for the training of the ablest men of the nation, within their domain. They did not, however, understand it at first. The movement began outside the colleges, and developed, in a certain measure, in spite of them. The engineering schools and the schools of agriculture are closely associated in their history, from this point of view.

The oldest engineering school in the United States is the *Rensselaer Polytechnic Institute* at Troy, in New York state, founded in 1824, with a very remarkable program, for its time, and which even today is very prosperous. About 1850, as has been said, Harvard established courses in pure and applied sciences, which formed the *Lawrence Scientific School* (where Louis Agassiz found a chair), and Yale similarly organized the *Sheffield Scientific School*. These two schools have remained, in a certain measure, distinct from the college. Harvard's has undergone rather numerous vicissitudes, and has for the time being, in a very large measure merged, so far as teaching is concerned, with the Massachusetts Institute of Technology. Yale's still exists, and has nearly 800 students, who are not mingled at all with those of the college proper. The degree of

Bachelor of Science (S.B.), has never had a prestige entirely equivalent to the A.B.

In a general way, the colleges, imbued with an un-yielding traditional classicism, up to the middle of the nineteenth century did not show any eagerness to favor the development of the applied sciences. On the other hand, the country felt keenly the need of scientific edu-cation, and in a very utilitarian form. This conflict resulted, in 1862, in the midst of the Civil War, with the passage by Congress of the Morrill Act, which has been of capital importance in the history of technical education and even of the universities in general. By the terms of this law, each state or territory of the Union was given as many times thirty thousand acres of public lands as the state had representatives and senators in the Congress. Thus the populous states of the East received large tracts of land, about a million acres for New York state, 780,000 acres for Pennsyl-vania, etc. These lands could be conveyed. The pro-ceeds must be devoted to education, and in preference to the teaching of agriculture and mechanic arts, but without excluding classical education.

The text of the law provided for "the endowment and maintenance of at least one college, whose principal object shall be, without excluding other scientific and classical studies, and including military instruction,[1] to teach the branches of knowledge in relation to agricul-ture and the mechanic arts, under the conditions which the legislatures may respectively prescribe, in order to develop the liberal and practical instruction of the in-dustrial classes, in view of the various enterprises and professions of life."

[1] It was the time of the Civil War.

The resources provided by this law were applied, in each state, to the creation of an institution which generally took and often still possesses the title of *Agricultural and Mechanical College*. I shall return to these establishments when I consider the agricultural schools. For the moment, I limit myself to recalling that, in many cases, they were the first nucleus of the present state universities.[1] Consequently, the presence of an engineering school in the latter is natural and in some sort congenial.

But today, almost all the large or medium sized universities, whatever their origin, have one. And besides, there are a certain number of independent schools of technology or polytechnic schools, some of them very important. I shall mention, for example, the following:

Location	Name	Founded	Number of Students
Boston, Mass.	Mass. Institute of Technology [2]	1865	1700
Brooklyn, N. Y.	Polytechnic Institute of Brooklyn	1854	786
Chicago, Ill.	Armour Institute of Technology	1893	527
Cleveland, Ohio	Case School of Applied Science	1880	534
Hoboken, N. J.	Stevens Institute of Technology	1871	324
Pittsburgh, Pa.	Carnegie Institute of Technology	1905	1219 [3]
Troy, N. Y.	Rensselaer Polytechnic Institute	1824	626
Worcester, Mass.	Worcester Polytechnic Institute	1868	535

I have not had at hand recent and complete statistics of the engineering students. Statistics given by *Science*,

[1] That is the case with the following: Arizona, Arkansas, California, Florida, Idaho, Illinois, Indiana (Purdue University), Kentucky, Louisiana, Minnesota, Missouri, Nebraska, Nevada, Ohio, Tennessee, Wisconsin, Wyoming.

[2] This institution has shared in the benefits of the Morrill Act.

[3] Of these, two hundred and sixty-eight are women.

in 1909, showed 144 technological schools or engineering colleges. One hundred of these schools, in 1907, already represented more than 33,000 students. On the other hand, here are some figures which show the importance of the engineering schools in some of the large universities in 1915–16:

	Students		Students
Michigan	1498	California	712
Cornell	1437	Pennsylvania	611
Purdue	1400	Missouri	564
Illinois	1039	Cincinnati	474
Ohio	841	Stanford	434
Wisconsin	758	Harvard	422
Yale (Sheff. Sch.)	790	Columbia	341 [1]

These indications suffice to show that a real army of engineers is constantly being trained in the United States, and in large part, in the universities.

The instruction covers many and various specializations, according to the region and its peculiar needs. The engineer's education looks less toward giving advanced scientific knowledge, and more toward a practical preparation. The principal divisions of the engineering schools bear the following names: civil, sanitary, mechanical, electrical, chemical, mining engineers, and metallurgists. These are the most important, and are in all the schools. But in certain schools there is a special section for drainage and tiling engineers. In the southern states, like Louisiana, there are sections for sugar mill engineers and for engineers of the textile industries (Georgia, North and South Carolina, Texas). In California and in the states using dry farming (Utah, Wyoming) there is a special section for irrigation. There are special sections for naval architecture (Massachu-

[1] As one sees, the large classical universities are not at the head as to engineering schools, at least in number of students.

setts Institute of Technology, Michigan), for aëronautics and aviation (Massachusetts Institute of Technology). The Armour Institute of Chicago has special courses on fire protection engineering.

The level of studies varies with the institution, but in a general way it is relatively low. Harvard, faithful to its general system for professional schools, had of late years sought to make a graduate school of its school of applied sciences (Lawrence Scientific School), but has given it up.

Today the plan of engineering studies is modeled after that of the classical college — four years lead to the degree of Bachelor of science in engineering. In the course of the classical college studies, one can pass into the engineering schools, and the studies already taken are given credit. In normal conditions, the first year is common to all the specialties, and includes the elements of the sciences, design, and the study of modern languages. Specialization begins in the second year, which still includes much instruction in pure sciences. The technical courses are largely placed in the third and fourth years.[1] Each section includes many special courses, some required, others optional, and these options are extremely varied. The year's instruction is completed by several weeks spent in camps, in practical field-work, during vacations.

Harvard, for example, as has already been said, has an engineering camp of 750 acres in New Hampshire, where for eleven weeks each year, exercises in surveying, topography and laying out railroad lines are car-

[1] The ordinary studies may be prolonged and deepened during a fifth year, leading to the degree of Master of Engineering. In certain universities, in particular at Harvard, there is even a degree of Doctor of Engineering, given on conditions parallel to the Ph.D.

ried on, and a mining camp in Vermont where for six weeks the students may practise underground exploration, the management of the machinery, and the diverse operations which the engineer is called upon to carry out in the field.

I have not the necessary qualifications for dealing thoroughly with the engineering schools, but there are a certain number of facts which seem to come out clearly enough to be set down here.

The first is the immense development of this kind of teaching and its direct connection with the industrial activity of the country. However, the universities are directly associated with this movement, while with us they are outside of it. We often speak of the necessary relations of science and industry. One of the conditions which can develop them is to interest directly the scientific centres in the training of the industrial personnel. And at the same time it is an important question for the vitality of the universities. If you take from them *a priori* all those youths who look at things from the industrial point of view, you weaken them in an almost fatal way. On the other hand, industry is on the border of pure science. Between the pure and the applied sciences, a barrier is erected which does not exist in the nature of things, and which would not seem to exist if theories and applications rubbed elbows in the same schools.

A second fact is the breadth of the modern equipment of the engineering schools and technological institutes. In 1916 the Massachusetts Institute of Technology of Boston moved into new buildings on Charles River. They cover more than fifty acres. I must note how numerous and vast the laboratories are: special laboratories for steam and compressed air, for hydraulics, cold, tests of materials, gas motors, measures of force, for mines and metallurgy, physics, chemistry, physical chemistry, applied chemistry, electricity, biology and public health, bacteriology, geology and mineralogy, and aërodynamics. Each of these laboratories has powerful machines which are not toys, not to be enumerated here.[1] The buildings, which have just been finished, cost no less than $3,500,000, in large part given by an anonymous benefactor. The land cost a million dollars. The equipment is estimated at $750,-000. The complete program of reinstallation comes to $7,000,000. Such is the scale on which a great engineering school is rebuilt in America today!

A last remark which I permit myself is that the conditions of training of the American engineer and of his French colleague are very different. The latter certainly has a very marked superiority for theoretical scientific instruction. I was told moreover, that since the war has brought into the American factories a rather large number of our engineers, the fact is perfectly recognized. There is nothing in the United States comparable to the preparation in our courses of the *École polytechnique* or the *École Centrale*. The first-year students, the freshmen, of the engineering schools, are very weak.[2] It is none the less true that the Ameri-

[1] Cf. *Bulletin Massachusetts Institute of Technology*, vol. lii, 1916, pp. 353 ff.

[2] Mr. R. C. Mann, in an investigation published by the *Bulletin of the Society for the promotion of Engineers' Education* (vol. viii, 1916) gives the results of tests made of the freshmen in 22 engineering schools. For example, only a third of them could calculate exactly, for $x = \dfrac{a+b}{2}$ the value of the algebraic expression, $\dfrac{(x-a)^3}{x-b} - \dfrac{x-2a+b}{x+a-2b}$.

can engineer gives abundant proof of all the qualities which are expected of him. What is asked of him is "not to be a savant, but a practical man, a business man and a financier. His art is not only to adapt the forces of nature to the use of man, but to do it economically. . . . The engineer must not build a fine bridge, with costly details, difficult to execute, in the desire of leaving a monument behind him." [1] He is first of all a man of action.

The difference between applied science and pure science is not in the methods (accordingly bringing the two together in the university is good), but in the end, that of the first being utilitarian, and that of the second philosophical. The practical has its own value and dignity, but it must rest not on empiricism but on a scientific basis. The engineer's education must be inspired with scientific principles, but not lose sight of the practical side. It must not be theoretical, but its motto must be, as Mr. McLaurin, the president of the Boston Institute of Technology said, "learning by doing."

This practical character is the fundamental trait of the training of American engineers. It is sometimes pushed very far. At the University of Cincinnati, the engineering students work in alternate periods in the university laboratories and in the factories of the city, with which an arrangement has been made to this effect.

The truth would probably lie between our system and that of the Americans. The latter would gain by having engineers with a more solid scientific instruction at the foundation — in that as in the other parts of the univer-

[1] Swain, *Science*, January 2, 1910, pp. 81–93.

sity, the real problem is the strengthening of secondary studies — but the education of our engineers is much too theoretical, not even useful in real life, and turns the mind away from the practical conception of things. Think of the mathematical education of the Polytechnic students, and even their education in physics and chemistry. What share has the laboratory — and real life — in it?

On the other hand, the American engineer's career is determined by what he has to give in life. The diploma with which he starts out plays, so to speak, no part. He is judged by his acts as a mature man, not on a prize won in youth at school, under conditions which have no relation to those which make for the man's worth. They do not begin by eliminating, by way of prizes, the greater number of the young men, while giving to a minority the advantage of a formidable handicap, which often turns them away from all serious effort on the day when it ought to begin, and which makes them believe in a definitive superiority, before it has been put to the test of life. The American approaches life at twenty-two, without being tired out by the conventional school work, without being spoiled by the success he may have had in it, or discouraged, but with the feeling that life is just beginning. The Frenchman of that age, often with his head buzzing with theory, is already tired and has the illusion that he has finally stood the test.

11. *Agricultural Schools*

The Morrill Act of 1862, completed by other legislative acts which have added new gifts, was the point of departure for very widespread teaching of agriculture, as well as of engineering. Agriculture was, before industry, and is today as much as the latter, one of the fundamental resources of the United States. Its conditions are very different from those of Europe, and resemble those of industry. The enormity of distances, the labor difficulties, the biological conditions, often very different from ours, have compelled great innovations, which were all the easier because there was, on American soil, no obstacle of traditions, more than a thousand years old. The value of scientific methods is today understood everywhere by the American farmer, and the diffusion of agricultural instruction, through the universities and the Agricultural and Mechanical Colleges, has a great deal to do with it.[1] The farmer of the younger generation has passed through one of them; he has the idea of the power of science and of method. This mentality explains the rapid propagation of the processes of dry farming and of irrigation. The immense and magnificent orchards of California suggest the psychology of an industrial environment much more than that of a farming environment.

The teaching of agricultural biology is scarcely represented in the eastern universities. Yet Harvard had one of the first agricultural schools, but it has now transformed it into an institute of general applied biology, devoted especially to the experimental study of

---

[1] A great farm near Chicago, such as one I had an opportunity to visit, reveals a quite different kind of life and of methods from those of our rural districts. In spite of the remoteness from urban centres, and the isolation, it is much more impregnated with the city atmosphere and ideas. It is true that the distances are today very much lessened by the automobile. In Kansas, a great agricultural state, there was in 1916 an automobile for every five inhabitants. That is to say, there was scarcely a farmer who did not own one.

heredity, the Bussey Institution. But it is especially the universities which have benefited from the Morrill Act which show a great development toward agriculture, and in which it has a special college.

Cornell University, at Ithaca, N. Y., at the time of its foundation, received lands granted by the Morrill Act to New York state — more than 1,000,000 acres. Its college of agriculture is the most developed of all, and has more than 1500 students. It is housed and equipped in a very complete and modern manner; outside of the fundamental scientific courses, a very complete group of special courses adapted to agriculture: plant physiology, horticulture, pomology, vegetable pathology, growth of plants, entomology, animal physiology, biological chemistry, forest biology, structure of soils, rural economy, farm equipment and management, agricultural mechanics, stock and poultry raising, dairying, etc.

Entomology, in particular, under the direction of Professor Comstock, for over forty years, has had a development which it has attained nowhere else, and Cornell is one of the principal centres for the training of the staff of the Federal Bureau of Entomology, which will be considered in the second part. There are no less than twenty courses in entomology, general and specialized, elementary or advanced. The studies in entomology can be combined with those in agriculture and in botany. For these sciences there is as rich a diversity of studies as for the classical college. A detailed description of them will be found in Mr. Paul Marchal's book, which I have already had occasion to cite.[1]

The Illinois and California universities, and even a

[1] *Loc. cit.*, pp. 250-287.

certain number of others, like that of Nevada, also have large colleges of agriculture.

Here are the numbers of students in the agricultural colleges of a few universities, in 1915–16:

Cornell	1535	California	540
Wisconsin	1091	Minnesota	598
Illinois	958	Missouri	536
Ohio	973	Nebraska	436

At the University of California there are numerous courses relating to the various branches of agriculture in that state, such as courses in oenology, citriculture, pomology, and oleiculture. The instruction in the college of agriculture is combined with that of the college of engineering, for example, so far as irrigation is concerned.

The plan of the studies of the agricultural colleges is modeled on the classical college, four years leading to the degree of Bachelor of Science in agriculture. There are also higher studies leading to the degrees of Master and even of Doctor. In 1913–14 8503 Bachelor's degrees and 1497 higher degrees were granted, including the Agricultural and Mechanical Colleges. Of the Bachelors, 1903 were students in the agricultural courses.

The state universities and their agricultural instruction cannot be isolated from the agricultural colleges, of which they are an extension.[1] Universities and colleges, the progeny of the Morrill Act today number sixty-nine. Seventeen of this number, exclusively for negroes, are of a primary level, and are really

[1] Here is the list of the courses given in these colleges: agriculture, horticulture, forest biology, veterinary science, engineering (mechanical, civil, electrical, mining, chemical, railroad, textile industry, etc.), architecture, domestic economy, chemistry, pharmacy, general sciences.

workshops. But the following figures, relative to all, show what enormous resources are dedicated to the diffusion of agricultural and mechanical knowledge, and how these resources have grown recently.[1]

	1892	1914
Number of colleges.....................	60	69
Number of volumes in their libraries......	724,000	4,395,000
Total value of their property............	$7,012,000	$60,298,000
Total revenues........................	$4,033,000	$34,891,224
Number of students in the colleges proper .	10,719	38,971 [2]

A small number of these colleges are specially agricultural, such as that of Massachusetts, at Amherst, which has played a part similar to that of Cornell, in entomology.

Some of them are very large. Those of Colorado, Iowa, Michigan, Oklahoma, North Dakota, Pennsylvania, and Utah, have several thousand students. They are interesting as still representing what the state universities were at the beginning, and from which a certain number still differ only relatively little.

12. Veterinary Schools

In 1913 the veterinary schools numbered 22, with 364 professors and 2481 students (only one woman). Veterinary instruction has only very recently developed

[1] The Morrill Act was completed by other laws, in 1883, 1890, and 1907. The two last, alone, carry an annual federal appropriation of $50,000 to each state. These laws have organized in every state, agricultural experiment stations, which are often in regular connection with the colleges or universities.

[2] Thirty-seven per cent follow the agricultural courses, 40 per cent the mechanical courses, 13 per cent those in sciences, 10 per cent those in domestic economy.

in a regular and methodical manner, and in the agricultural colleges or the state universities. The first course relative to the veterinary art was begun at Cornell in 1868. The oldest school dates from 1880. The most important is that of Cornell. There are also important ones at New York University, the University of Pennsylvania, Ohio State, and George Washington Universities. The course of veterinary studies is generally three years.

To conclude this rapid review of the schools, into which an American university of today is divided, I must add that their respective limits are not absolutely rigid. We have been able to see, in the course of the preceding survey, that, for example, the students of the schools or colleges of agriculture, of industry, of commerce, or of education, take the general cultural courses under the Faculty of Arts and Sciences. There are even many courses common to the colleges of agriculture and of engineering. A certain confusion may result from this. But on the other hand, there is an advantage in that the university keeps its unity and that the more and more numerous subdivisions which must necessarily be differentiated, are not separated by watertight bulkheads, as are, for example, our Faculties of Letters and of Sciences.

CHAPTER XI

UNIVERSITY EXTENSION AND THE SUMMER SESSION

Importance and character of the summer session. University of Chicago quarter system. Extension proper: its beginnings. Chautauqua institutes. Extension at Harvard, at Columbia, in the state universities (California, Wisconsin). Breadth of university extension.

THE activity of the American universities is not strictly limited to their regular instruction. It has a very considerable complement in university extension under its diverse forms, and in the summer session, which is a particular case of extension, but which we shall consider first.

Harvard seems to have been the first university to hold vacation courses, beginning with 1871. But especially of late years this practice has been generalized and attendance has been greatly broadened. Yet there are a certain number of universities, like Yale and Princeton, which have not adopted it. In order to give an idea of its success, it suffices to indicate the numbers enrolled [1] at this session, in a few centres, in 1915.

	Students		Students
Columbia	5590	Cornell	1436
Chicago	3984	Harvard	1250
California	3179	Illinois	938
Wisconsin	2602	Minnesota	867
Michigan	1594	Johns Hopkins	350

Attendance at these sessions is quite different from that of the academic year. It consists mostly of ma-

[1] These students do not figure in the totals given in the preceding chapters of this book as the regular population of the universities.

ture men and women — of whom a great many are graduates — who come to complete their instruction in a definite subject, or to bring it into the current of recent progress. Women are admitted, even in universities like Harvard, which exclude them from their regular studies. Physicians, engineers, teachers of secondary or primary schools, come to study a specialty. In general they enroll in a single course only, for which they pay a rather high fee, $10 to $60, according to the course, at Harvard. This course may include numerous meetings, and the session lasts, depending on the university, from six to eight weeks, in July and August. The ordinary students too are often allowed to enroll in one of these courses, and they can thus make up a deficiency in the current year, or hasten the end of their studies.

It is an extremely practical institution, which permits many classes of persons to complete their education without giving up their ordinary occupations. These auditors, more serious and more exacting than the ordinary students, have bid farewell to all the frivolities of the college. In certain respects, this summer session is nearer to higher education as we understand it, than that of the normal year.

The universities make many exchanges of professors for this session. Many members of the eastern universities go to Berkeley, for example.

The case of the University of Chicago is rather peculiar, and deserves to be specially noted. In reality, the University of Chicago has completely suppressed the vacation period. It works the year round without stop, and the term unit there is the quarter, instead of the half-year. The courses are consequently combined.

Each professor should have three quarters of work and one quarter of rest, which he can take at his choice, at one season or another. The vacation period here is simply the *summer quarter*, in which the university offers its usual resources. That is the cause of the special success of these vacation courses at the University of Chicago. Other universities should be tempted, it seems, to imitate this innovation, which moreover, allows professors to be free at periods other than the usual vacations.

The Summer School is only a very special case of university extension proper, a very democratic work, whose program is immense and generous, but not free from whims and sometimes from a demagogic spirit. In fact, it is a question of bringing science into contact with the people, of allowing men and women, already absorbed by professional occupations, access to culture and to the university degrees, and above all to make known the applications of science to the masses, so as to make these applications come into use and thus to hasten progress.

Before the universities themselves had undertaken this task, in which, however, the English universities had preceded them, different endeavors, organized mostly by university people, had begun it, in the form of popular lectures, correspondence work, and the organization of public debates.

Such was the American National Lyceum, founded in 1831, in which men like Daniel Webster and Emerson actively collaborated.

Such, above all, was the Chautauqua Literary and Scientific Circle (C. L. S. C.). Chautauqua is the Indian name of a lake in New York state, on whose shores, in 1874, an enormous work of popular instruction during the summer, was organized. A city was created there, which exists only during the few weeks of the session, and which then attracts more than 10,000 persons. Almost all the subjects of the college program, music, etc., are taught there. Aside from the regular courses, grouped in a cycle of four years, there are lectures, public debates, concerts, and dramatic performances. On this model numerous daughter institutions have been created, which bear the same name, and Chautauqua circuits, a group of lecturers, actors, musicians, from July to September, make a tour of a series of towns in which this system of teaching is applied, thus reaching a large public.[1] The name Chautauqua stands for advanced popular instruction.

Permanent foundations, like the Lowell Institute at Boston, and the Peabody Institute at Baltimore, are also for popular education through lectures and other means.

After various vicissitudes, extension work has been firmly entrenched and developed in the universities for twenty years past. The western universities have undertaken it on the vastest scale. A few years ago the University of Chicago had organized it at Chicago and elsewhere, and its lecturers radiated into twenty-eight states, addressing nearly 50,000 auditors. The old universities of the East are also taking part in this movement. Harvard, in collaboration with the neighboring institutions (Boston University, Massachusetts Institute of Technology, Tufts College, Wellesley College,

[1] See H. B. Adams, *Monograph on Education in the United States.* (Paris Exposition, 1900), Monograph no. 16. In it will be found a bibliography.

etc.), has created, at several points in Boston, regular
courses, parallel to those of the university, and leading
to a degree of *Associate in Arts, A.A.*, which may give
access to the Graduate School of Arts and Sciences, and
thus lead to the degree of Master.

Extension at Columbia shows a very large development, especially in the form of evening courses, in the
university buildings or in various places in New York.
The catalogue of the University for 1912–13, the most
recent I have been able to consult, mentions no less
than two hundred and fifty of these extension courses.
Thanks to them, it is possible, while busy at one's profession, to do the work equivalent to the two first years
of college, freshman and sophomore, bit by bit, and to
enter the university later in a regular manner to finish
the course for the Bachelor's degree. Besides, certain
of these courses have an essentially practical character:
thus the department of physics of the university gives
courses in optics for opticians.

In the state universities, extension fills an enormous
place. It is one of the means of justifying in the eyes of
the people, the huge expenditures made for higher
education, by bringing the knowledge which they can
assimilate or which may be useful to them, in contact
with the masses everywhere.

The forms which this extension takes are many.
There are lectures and even regular courses, in the chief
cities of the state, and even in unimportant centres.
To this end they bring together, in each of these cities,
in a permanent way, appropriate means of demonstration, cinematographs, projection lanterns, even actual
laboratories on a small scale, and also a nucleus of persons in settlement to aid the traveling lecturers. Dis-

cussions are also specially organized. And there is correspondence work. The applications of biology to agriculture are among the subjects which are the most abundantly represented, and that is explained by the importance of agriculture in the state universities and no less by the political influence of the farmers in the agricultural states of the Middle and Far West. They organize numerous demonstrations on the farms themselves; special trains travel over the state, carrying material and a staff. There is sometimes a rather demagogic stage-setting there, but apart from certain exaggerations, it remains none the less true that all the applications of the sciences can thus be brought into direct contact with the farmers, and that that contributes toward facilitating the application of new processes, and toward developing the taste and the feeling for progress in the rural population, and toward restraining very much their spirit of routine. Thanks to the prior development of these works, they have been able to make extremely powerful campaigns of social interest, for example, against alcoholism or tuberculosis.

The University of Wisconsin is one of those which have conceived this work on the vastest plan, hoping to spread in the whole community which surrounds it the spirit which animates it, and the practical results of science; to be itself in some sort present everywhere. Moreover, it receives from the state an annual subsidy of $30,000, for extension in the domain of agriculture alone.

The University of California has also made a great work of its extension, and one which it tries to spread afar into numerous cities. It has created within itself, for the methodical organization of this work, a special section under the name of *Department of University Extension*, which includes five bureaus: one for the organization of regular courses in various cities; another for correspondence work in the various sciences; a third for the organization of lectures; a fourth to organize public discussions, which acts especially through the distribution of bulletins, bibliographies, programs, etc.; finally, the fifth, called *Bureau of Municipal References*, popularizes all the questions of hygiene and urban organization by way of bulletins or inquiries. In 1910, of thirty-two state universities, twenty-three had organized extension, and fifteen had created a special department with this end, as we have just seen for the University of California.

The Chautauqua system has served as a general model for all these enterprises. You see how much breadth this extension work has, and what social usefulness it may possess; also how it draws closer together the university and society, science and the people. Still more than the existence of engineering or agricultural schools, it marks the utilitarian, realistic, and democratic tendency of the state universities in the West. In spite of the necessary imperfections in this work, which is still in its beginnings, it does open the minds of the masses and accelerates progress.

CHAPTER XII

GENERAL CONCLUSIONS ON THE ORGANIZATION
OF THE UNIVERSITIES. UNIVERSITIES
AND SOCIETY

Insufficiency of preparation by secondary education. Broad contact of the university with youth. Evolution of the universities. Rôle of the state universities. Broadening of the social function of the universities. Contact with society. Rôle of the alumni. Loyalty and donations. Links with the university: clubs.

AFTER having passed in review successively the diverse parts and the diverse modes of activity of the universities, it is fitting to cast a general glance over them and to bring out the most essential facts relative to their present state and the probable course of their further development. Institutions, as Mr. Eliot has justly said, are more interesting through their tendencies than through their immediate condition.

The notion of a university, in all the great countries, at present answers a double object: teaching of the higher branches of human knowledge and organization of original research, in order to push back still further the limits of our knowledge. By unanimous consent, it is this second mission which seems the more essential and that which is truly specific. The American university world constantly affirms this conviction. For the moment, however, I leave it one side, and shall return in the second part of this book to the examination of the American university from the point of view of research. I am now considering it only from the point of view of teaching. For after all that is the fundamental element. Research can be built soundly only on the foundation of solid instruction.

American universities have a very great power, in that they attract all the youth. All higher education is carried on within them. More and more the technical and professional schools tend to come back into them. Those which develop brilliantly outside, like the Massachusetts Institute of Technology, are, in fact, slightly specialized but still true universities. Young people enter them in the same way and leave under the same conditions. The fact to be emphasized is that the doors of all these establishments are wide open and that no one of them gives to those who leave it a monopoly for certain careers.

The universities have set themselves the task of furnishing, for all branches of social activity, the leaders whom a higher education ought to train. Nothing hampers them in this program. They extend it more and more, and having liberty and autonomy, free competition is, for them, an active stimulus to perfect its realization.

The great problem of teaching which they have to solve at present is to conciliate the necessity for general education, assuring breadth of views and culture, with that of the technical teaching required for the various careers. This problem arises in all countries. What are its difficulties and its special modalities in the United States?

The general training of the mind ought to be at least well prepared by secondary education. That was the virtue of our classical studies, and we ought to conserve it tenaciously, while taking account of the modifications which the general present conditions must cause their former arrangement to undergo. Secondary teaching seems to me to be the weakest point of the American system of education. The student who comes out of the high school at eighteen has not a sufficient intellectual training. A good part of his university studies consists in finishing his secondary studies. A number of the best qualified American educators, W. R. Harper, who was the first president of the University of Chicago and who put it brilliantly in the first rank, D. S. Jordan, who did the same for Leland Stanford, E. J. James, who is now President of the University of Illinois, and many others, recognize that in a general way the first two years of college ought to be put back in the high school. The real problem is at the same time to have the young Americans finish these studies at eighteen, as is the case in France and Germany. Four or five years in the university would then suffice to complete the theoretical education and give the technical education necessary for the various careers. The four years of college, from eighteen to twenty-two, a mere preparation for further technical studies, are evidently too long, and are a legacy of the past which cannot continue. Fundamentally, in the past, the college was simply secondary instruction.

From this earlier condition, the American university has kept, to its advantage, the habit of a close and methodical control over the work of its students. It treats them, in this respect, as boys, who must be followed attentively, not as mature men whom it can allow to act at their own will. This habit has been transmitted to all its new parts. The chief reproach which could be made to its teaching, in a general way, is that it is not sufficiently impregnated with synthesis. Mr. Woodrow Wilson made this criticism by declaring that we must not confound information and education. The American student is not left enough to himself, and led to reflect. He is constantly guided. But the theoretical and practical instruction offered him is very well coördinated, and when he really has the taste for work, he can draw excellent results from it.

One of the points which seem to me most important in the evolution of the American universities is the place which the applied sciences have taken in it, in particular, all that concerns engineering and agriculture. The universities have thereby escaped from the danger of a mandarinate. Institutions for higher education (I leave one side those which are completely specialized for research) do not seem to me really able to live, in modern society, on the basis of the speculative sciences alone. I do not at all wish to belittle the latter, and the university is their true and only home, but they need contact with reality to remain living.

It is sound that all speculation should be tempered by consideration of the real, and likewise that speculative teaching should be in the same surroundings with practical teaching. I believe, therefore, that an organization like that of the modern American college, which associates the pure and applied sciences, is in principle preferable to one which, like our own, isolates on the one hand faculties of science, and on the other, technical schools. This has the double advantage of not opposing pure science and applied science, and of not creating institutions which cannot really recruit their own numbers and which end fatally in a mandarinate system.

The university which is both theoretical and practical is a much more real representation of society.

Undoubtedly the Morrill Act was the great ferment of the development of technical and agricultural instruction in the United States. In the state universities, which sprang from it, this teaching at first took, and in many cases still has, a too radically utilitarian spirit, which political influences tend to impose. Little by little, however, this excess of utilitarianism is fated to give place to a broader conception. The existence and spirit of the private universities suffices to draw the state universities into the path of general culture. In his book on the American universities, Mr. Slosson justly notes the immediate and large influence which the creation of the University of Chicago, in 1890, exercised over the state university of Illinois, by bringing about in the latter a great development of the instruction in pure culture.

The dualism, and up to a certain point the rivalry, of the private universities and the state universities, seems to me an extremely favorable circumstance. The first have evidently implanted, and till now represented true intellectual culture in the United States, but if they had been alone, they would perhaps have been too narrowly confined within their classical tradition, and in spite of everything, in a too narrowly aristocratic form of education. Is not that, moreover, the story of Oxford and Cambridge up to a recent period? The existence of the state universities has undoubtedly driven them to broaden their field toward the modern needs of society. On the contrary, they are by their very qualities, the witness which obliges the raw and violently utilitarian democracies of the West to let their universities evolve toward culture, and to raise their standard. Under the influence of these two tendencies, the teaching of the applied sciences remains practical, and little by little its basic level is raised.

The philosopher Royce, so much esteemed by everyone at Harvard, a pure logician by profession, was certainly not of a spirit which could be taxed with limited utilitarianism. He has shown, moreover, from the first phases of the present war, what high idealistic sentiments animated him.[1] In 1909 at the Congress of the American Association for the Advancement of Science at Baltimore,[2] he characterized, in a very just and very profound way, in my opinion, the opposing tendencies which divided the opinions of American educators, the spirit of the old classical college, and that of the modern western universities. One cannot think, he says, of opposing radically what is called the college to the technical and professional studies. "One may protest as one will that one misuses the term college when one talks of a college of agriculture, and that one ought instead to speak of a technical school of training in agriculture. . . . But whatever one does by way of formulation, of definition, and of criticism, the state universities will continue to show that the best things that you can do for the young men who are to be trained in the humanities is to keep both them and their teachers in pretty close contact with the pupils and teachers who are engaged in technical studies. . . . For my part," he says, "I suppose one of the notable functions of an academic institution to be the uniting rather than

[1] See, in particular, his speech, "The Duty of Americans in the Present War," delivered at a meeting in Tremont Temple, Boston, in January, 1916.

[2] *Science*, March 12, 1909, pp. 401–407.

the further sundering of the various more or less learned activities of modern life, the humanizing of engineers, and the preparation of the young followers of the humanities for some practical service of mankind."

The eastern universities must, in the future, according to Royce, broaden their plan more and more, after the type of the state universities. "The centre of gravity of our future American academic life can not always, can not I think very long, remain east of the Alleghenies. Through a perfectly natural and inevitable evolution, the state universities of the Middle West and of the Far West, supported as they are, and will be, by the vast resources of the communities from which they emanate, and guided by an educational ideal ever being perfected, will occupy, in one or two generations, an almost central place in American academic life."

The universities are, therefore, according to this authoritative forecast, definitely committed to that path, in which their rôle, as Mr. E. J. James, president of the University of Illinois says, is "to provide for the training of the youth of the country for all the careers requiring an extended scientific preparation, based on an appropriate liberal education."[1] They will establish new specialized colleges for new needs. "Any profession can be practised rightly only on a scientific basis." Therefore, in brief, positive science becomes the basis of preparation for practical life and inspires all the activity of the university. The university is to diffuse this spirit of positive science into all the divisions of society.

This movement dates from yesterday; yet it is being accomplished in the state universities with a speed of

[1] *Science.*

realization which is part of the American temperament, but which is perhaps not yet sufficiently marked by calmness. "These universities," says Mr. J. M. Baldwin,[1] "are the field on which all sorts of pedagogical experiments battle, where the newest and boldest theories are put in practice, and where 'up-to-date' methods receive an often premature application. They seek constantly to obtain practical results, which may impress the exacting public which pays the taxes. Hence there is a veritable whirlpool of ideas and methods. A state of mind characterized by the urgent need of action, but which at the same time lacks assurance and confidence, is produced." It is to be hoped, however, that equilibrium will be established little by little. In any case, by turning, in a fashion perhaps at present excessive, in a utilitarian direction, the universities are but returning to the tendencies of one of the founders of American society, whose idealistic intent, at the same time, we cannot deny — Benjamin Franklin.

The tradition of the old private universities of the East on the one hand, the radical and utilitarian spirit of the state universities on the other, are the two antagonistic elements between which we must hope to see established a compromise which will maintain the rights of culture. This result would be much more surely gained if the student arrived at the university already better trained and more cultivated.

The universities have another solid contact with society, one of a traditional and sentimental order, and in fact of rather aristocratic tendencies. It is the at-

[1] *Foi et Vie, Cahier B*, 1917, p. 15.

tachment which links every American to the institution, college or university, through which he has passed. This loyalty is a characteristic of their way of life in general, but it has a special importance for the private universities, for on it, in fact, their whole existence is based. Its force and prevalence are one of the undeniable marks of an idealistic side in the American mentality. And of course the universities are carefully on the watch to maintain it. It rests on the solidarity and comradeship which college life establishes between the students, and which in some degree identifies their memories of youth in an agreeable form with the institution through which they have passed.

The university becomes the centre of a vast family, so much the more powerful, the more numerous it is.[1] It deserves the name of *Alma Mater*, and its foster children, its alumni, consider it a duty to provide for its needs, after having been educated by it. Gifts to universities have thus become a normal element of the civic activity of the wealthy class. They suffice to assure, not only their existence, but their development, and often even with an excessive luxury. They permit vast conceptions and rapid realizations to those who hold in their hands the destinies of a university. Examples abound.

At Princeton, my colleague W. B. Scott, the eminent paleontologist, walking with me across the campus, showed me with pride the seventy-five large buildings which stand there, magnificent laboratories, sumptuous halls, dormitories, all built with gifts of alumni.

When Harvard built its magnificent medical school in Boston, a considerable sum was lacking to erect one

[1] Cf. Table p. 269, col. 6.

of the five large buildings which compose it. They went and explained the situation to the banker, Pierpont Morgan, who, after having listened and reflected, replied simply, "All right, sirs," and promised the sum. It was a matter of more than a million dollars. Those are solutions which have nothing of bureaucratic slowness and red tape.

In April, 1912, a young graduate of Harvard, Harry Elkins Widener, perished on the *Titanic*, at the same time with his father. His mother, who escaped drowning, gave the university the collection of books which her son, an ardent bibliophile, had gathered. The university was at that time planning to rebuild its library, which was too small for the six or seven hundred thousand volumes it contained, and above all, absolutely insufficient for the future. Mrs. Widener easily allowed herself to be persuaded to associate the memory of her son with this reconstruction. She took entire charge of it; her architect executed the monument, on land designated and according to indications furnished by the university. The latter did not even know — at least officially — what it cost (it is said between $2,000,000 and $3,000,000). The cornerstone was laid in June, 1913. The library was dedicated in June, 1915, at Commencement, and completely installed for the opening of the following academic year. There also, no administrative formality intervened to trammel the gift nor to retard the execution.

Quite near Boston, Tufts College is an institution of moderate importance, whose buildings rise on the slopes and summit of a grassy hill, the view from which is magnificent. The biological laboratories occupy a building constructed with funds given by an alumnus whose career was not precisely intellectual, Barnum, the proprietor of the famous circus.

And these examples could be multiplied. There is scarcely a week in which *Science* does not record one or several important donations. It is one of the most usual ways of perpetuating a memory.

These last months too, we have had a touching example of it, and one which is very dear to us. Before America became our ally, more than one university was represented on our front by numerous alumni, Harvard by several hundreds. Last spring more than thirty of these young Americans had already fallen gloriously. Among them the aviator Victor Emmanuel Chapman was killed, June 23, 1916, at Verdun, in an aërial combat. To perpetuate his memory, a group of subscribers have founded a fellowship at Harvard, in his name, which will be awarded to a French student.

If you wish to appreciate the breadth which these gifts take, and what a factor they are in the development of the universities, you have only to consult the Report of the Commissioner of Education. Here are some figures taken from that for 1913–14.

The total of gifts made to the universities and colleges during that year, and coming to the knowledge of the federal bureau of education, reaches $29,927,137, and that is not an exceptional figure, for the total for the years 1901–14 is over $300,000,000.

The table on page 147 indicates, in dollars, the figures for a few universities.

These figures are evidently rather variable from one year to another. But they are always considerable. I will add that the gifts received in 1913–14 exceeded $100,000 in forty-five universities, and colleges, and the total for these establishments was more than $20,000,000.

University	Total Receipts	Receipts from Tuition	Income from Endowment	Gifts			
				For Current Expenses	For New Equipment	For Capital	Total Gifts
Harvard	$4,287,185	$895,497	$1,344,904	$256,239	$253,914	$1,379,356	$1,889,509
Yale	2,600,619	742,510	809,171	138,390	125,000	756,457	1,019,847
Columbia	6,685,869	1,017,137	1,138,875	468,607	114,936	680,647	1,264,190
Chicago	3,332,151	743,598	1,082,514	27,966	665,211	626,803	1,519,986
Cornell	6,790,260	535,346	610,208	8,623	3,000	4,364,486	4,376,103
J. Hopkins	738,049	121,130	244,210	19,420	10,681	118,909	149,010

The universities are evidently watchful to maintain the bonds which unite them to their alumni. They interest them in their life by giving them, as has been seen, an important part in their government. In most cases, indeed, the trustees are elected by the alumni. The ceremonies which end the academic year are an occasion to bring back a large number of them to the campus, and to awaken their interest not only for the university as they knew it, but as it is changing.

There are anniversary dates on which the tradition of the return of the classes is particularly observed, for example, twenty or twenty-five years after graduation. And these rites bring with them a gift to the university. At Harvard it is now a formal rule that at the twenty-fifth anniversary of graduation each class gives to the *Alma Mater* a sum of $100,000, which thus becomes an item of the ordinary budget.

University solidarity borrows from American customs another more constant and no less solid link, the club, which is the most living, and perhaps the most general form of association of American life.

We have already seen the perhaps exaggerated rôle which clubs play in the life of the student. Did we not go so far as to say that the college itself, especially where its traditions have been best preserved, was only a country club, where one spent four as agreeable years as one could? [1]

But it is through the clubs of former students that each establishment maintains and consolidates its family of alumni. Thus there are Harvard clubs in all the great centres of America, and even wherever there is a very small group of Harvard men. Honolulu has one, Paris also; the old American university gives us this example of solidarity and fidelity. In New York and in Boston, where the Harvard men are numerous, these clubs have each four or five thousand members, and have been able to house themselves in a comfortable residence, an animated and complete centre of Harvard life. Yale, Princeton, Cornell, likewise have their individual clubs in New York. The Massachusetts Institute of Technology of Boston has its own there too.

In the majority of cases, the clubs of the various universities in the same city, are federated in a general University Club, in order to have a luxurious material equipment, each one cultivating separately there its own memories.

Thanks to these clubs, there is scarcely a striking event in the career of the university in which its alumni, even the most distant, are not associated, often in a very direct manner. We had a particularly significant ex-

---

[1] The club also cements solidarity in the life of the professors, who meet, if only at lunch time, in a Faculty Club, present, under one or another name, in all the universities. All those who have taught at Harvard keep a pleasant memory of the Colonial Club.

ample of it in June, 1916, when the Massachusetts Institute of Technology celebrated its transfer to its new and magnificent buildings on the bank of Charles River. There were festivities of many kinds, to which each of the classes contributed its individual manifestation, and they ended, as usual, with a banquet in Boston, of which more than 1,500 partook. In thirty-four cities of the United States, from New York to New Orleans, and to the great centres on the Pacific coast, Los Angeles, San Francisco, Seattle, at the same hour, the Tech Clubs were also gathered at banquets. At the hour for the toasts, the banqueters in all these cities could have the illusion of being at the Boston celebration itself. With a telephone receiver at the ear, they were actually able to listen to the speeches delivered there and in Boston likewise, each of the banqueters could hear the greetings which were sent one after another from the various cities. The solidarity of the alumni, in this gathering of engineers, utilized the most modern means of manifesting itself.

And lastly periodicals and reviews, *Alumni Bulletin*, *Alumni Weekly*, *Graduates' Magazine*, etc., at regular intervals, remind the alumni individually of university affairs, and keep them in the current of all the great or small events which concern them; put them in touch with the plans formed, the material needs; in a certain measure submit these projects for their approval and at the same time ask them for the means for their realization. That is a heritage of English customs, and an important employment of private initiative, to which we cannot refuse our sincere and admiring approbation.

The alumni, then, bring enormous support to the universities, and at the same time exercise an undeniable influence over them. That does not mean that this influence is always beneficent. In their affection for *Alma Mater*, preoccupations of an intellectual order are not the most active. The mass of the alumni, especially the most of those who can make sumptuous gifts, are not scholars, and the memories of their college life are chiefly those which made these years a "good time," for them. It is the joyous, sporting, and worldly side of college life whose tradition the alumni are anxious to maintain. The university must compromise more or less with these tendencies, and devote a part of the resources which come to it, to increasing the luxury and agreeableness of the college, before thinking of the scientific needs. The universities which, like Johns Hopkins, have for themselves only the austerity of the intellectual task, do not attract a crowd of generous alumni.

That is nothing more than purely human, and the fact is that the universities still find easily the means of realizing their most strictly scientific *desiderata*, either among their alumni, or among very rich men, who have no debt of gratitude toward them. One can find no more noble use of a fortune than to devote it, as Leland Stanford did, to founding a great university, in memory of his son. Andrew Carnegie and J. D. Rockefeller, merely to mention their names, and one might add many others, figure among the generous benefactors of numerous universities. Mr. Carnegie has been guided in all his largesses, by an undoubted and ardent desire to contribute, through the progress of education in all its stages, to the amelioration of human conditions. Mr. Rockefeller, in 1910, making a last donation [1] to

[1] Mr. Rockefeller's gifts to this university have reached in all, $25,000,000.

the University of Chicago, of which he was the principal founder, announced at the same time that he was withdrawing his representatives from the council of her trustees, and he added: "I am acting from an initial and lasting conviction that this great institution, being the property of the people, should be controlled, conducted and sustained by them. I have merely had the privilege of coöperating in the generous efforts made for its building." The council of trustees, in accepting this last gift, declared that Mr. Rockefeller had never sought to use his influence, that he had never intervened for the nomination, promotion or recall of the professors, and that he had never made representations in respect to views expressed by them, even on religious questions, on which doctrines in formal opposition with his well-known views had been formulated.

It may have happened that donors have sometimes applied a certain pressure to the universities. One may regret, not without reason, that individual wealth should be able to exercise so great an influence. But in my opinion it would be unjust to deny to this great movement of liberalities, from which American higher education profits, a broad foundation of idealism and public spirit. After the account is cast up, we must sincerely admire it, and consider very fortunate those customs which interest and associate in the life and management of the university, all those who have passed through its ranks, or whom fortune has favored.

PART II

SCIENTIFIC RESEARCH

CHAPTER XIII

SCIENTIFIC RESEARCH IN THE UNIVERSITY

Its conditions. Selection of the personnel, and the sciences. Mr. J. McK. Cattell's statistics and the distribution of the best American scientists. The scientific equipment: laboratories and libraries. The relation of research and teaching.

THE first part of this book has shown us the American universities under extremely varied aspects, yet there is one which we have barely touched upon. That is the one which it is everywhere agreed to consider essential, scientific research. We shall consider it now.

And first, American intellectuals, especially scientific men, but also engineers, unanimously proclaim that of all the aims of the university, this is the supreme aim. The universities must, before all else, cause science to progress. "Research is the *nervous system* of the university," said Professor C. M. Coulter of Chicago, in a toast which I had the pleasure of hearing, April 15, 1916, at the banquet of the Philosophical Society. "It stimulates and dominates every other function. It makes the atmosphere of the university, even in the undergraduate division, differ from that of a college. It affects the whole attitude toward subjects and toward life. This devotion, not merely to the acquisition of knowledge, but chiefly to the advancement of knowledge for its own sake, is the peculiar possession of universities. . . . There must be an increasing determination to permit no other function to diminish its opportunity, and to allow no method of administration to depress its spirit." [1]

Research is, then, undoubtedly the ideal of the teaching staff of the American universities, and we are to examine in what measure it is realized. We have seen how vast and complex these universities are, and to how many divergent needs and traditions they respond. They are quite evidently not planned for research, which has made a place for itself in them recently. Is it favored or hindered by the general surroundings? Voices calling for a better adaptation are not lacking.

They regret the large place which the college and college spirit still have. The professors are overburdened with courses, too much absorbed by pedagogical preoccupations and the routine work which the students give them. Neither sufficient freedom of mind, nor time, remains for them to devote themselves calmly to serious researches. The teaching itself undergoes the influence of the inferior level on which the students are who come to the university. The college life weighs too heavily on the university. That is what Mr. D. S. Jordan, former president of Leland Stanford, expressed in a striking manner, in an address given at Yale,[2] in contrasting Yale College with Yale University. "We must," he said, "choose between the two conceptions: one, that of the college, a school for boys, with its football team, its glee club, and its crews; the other, that of the university, a school for men; and come out of the present transitional state. The glory of Yale, until now, has been Yale College; that of the future must be Yale University; but the two things in the same yard,

[1] *Science*, June 9, 1916, pp. 810–812.

[2] *Science*, March 19, 1909.

with the same teachers, the same discipline, this condition can never be a finality."

"The American university," says, moreover, Mr. A. G. Mayer, "today remains a hypertrophied college, and the conservation of the past is its ideal, rather than the revelation of the new truth. The professor in it is more and more overwhelmed by the pedagogical work. Since 1880 the universities have experienced an enormous material development, but disproportionate to their intellectual development. Large buildings and fine lawns may be necessary and are certainly desirable, but a university consists, first of all, in a staff of eminent professors."

"The American university," Mr. Schurman, the president of Cornell, says, further, from the same point of view, "is still in the state of expectancy or of promise. Its future is to be a great school of research."

On the other hand, the experience of Johns Hopkins and of Clark University, show the almost insurmountable difficulties in establishing, *independently of the state*, a university which shall be exclusively a school of advanced studies, and the American democracy does not yet like to subsidize institutions not having, at least in large part, immediate usefulness.

It is therefore very certain that the present situation carries much that is unfavorable, but we must not fail to recognize the real advantages: first, that solid foundation in society which it gives to the university, whether through the college traditions and the active sympathies of the wealthy classes, or through the development of the university toward applied teaching and contact with all the realities of modern life. A university which is entirely devoted to pure science is isolated in its tendencies from the surrounding world, and does not enroll enough students.

Pure science, and especially research, can be the work of only a small number of superior and disinterested intellects; these can be recruited with certainty only by a very wide selection. This selection gives good results when it is operated on large masses of individuals; it works badly if one operates on only a small number, as is the case every time that a faculty has limited itself to sciences purely speculative and without applications.

Therefore, without giving the impression that it is perfect, I believe that in principle the actual constitution of the American university, is not bad. It offers a very broad foundation, on which by working properly, you are in excellent condition for selecting the chosen few who will cause our knowledge to progress. What I personally saw at Harvard confirms me in this opinion. Evidently the selection is not easy to make, and one does not readily hit upon men of genius. "The making of a Darwin" is the title which Mr. D. S. Jordan gives to one of his presidential addresses to the American Association for the Advancement of Science,[1] in which he proves, in brief, that the universities of his country do not yet have the best recipe. For men of genius, the only useful and practical recipe, which is not too ambitious, is that the conditions of the environment shall not stifle them automatically. Systems of education should avoid this major defect, and for the rest, limit themselves to getting the best results from the average.

In brief, moreover, in thirty years, which is a short

[1] *Science*, December 30, 1910, pp. 927–942.

time, the American universities have realized, from the purely scientific point of view, enormous progress. The number of doctorates, if it is not a datum of an absolutely decisive value, is nevertheless an important indication.[1] Apprenticeship in research, through the doctorate, seems to me equally satisfactory. Evidently, as Mr. Castle observes, the fabrication of theses furnishes only a rather feeble result for the general progress of science, but even there, selection continues to operate, and can give only at rare intervals a really superior subject.

It is rather through the examination of the personnel that one can appreciate the value of the universities from the scientific point of view. And it is not doubtful that this personnel, on the whole, makes a big effort toward research; that in thirty years it has improved enormously; and that today there exist several great scientific centres full of vitality and independent of one another. There are six hundred colleges; there cannot be so many centres of discoveries. Only a very small number can be set up. By placing oneself at the point of view of research, one can rather easily determine the most important universities. They are, moreover, those which are in a general way the most prosperous.

The absolute autonomy of the universities, the material interest which they have in possessing as distinguished a staff as possible, the freedom which the presidents enjoy in supplying it, result in the selection being made, for the most powerful universities, at least in a large measure, according to the real value of the individuals, and in this selection, scientific works come in for a rather large share. Competition between the

[1] Cf. tables, pp. 96 and 97.

universities causes them automatically to hunt for the right man for the right place.

Mr. J. McK. Cattell, professor of psychology at Columbia, in his review, *Science* — which very faithfully reflects American university life, and particularly the scientific life — has published an interesting series of studies on the university teaching profession, which are distinguished by a democratic and very independent spirit. He has tried to apply statistical methods — perhaps sometimes with too much excess in detail — to the appraisal of individual merit, and to draw from the results obtained, judgments on the university environment and indications concerning the reforms to be applied to it. He has thus sought to determine the thousand most distinguished scientists of the United States,[1] and, these once known, to deduce a series of conclusions from their rise, from their distribution in the various universities, from the conditions of their career, etc. He puts into this thousand a number of representatives of each science proportional to the number of workers in that science.[2] For each science, he has asked ten leading representatives, specialists of authority, to classify the workers in their specialty in order of merit. This plebiscite related to 2481 names. In each science, the lists furnished have been combined in a general list, according to the averages, and applying the correction for error to the interpretation of the results.

Of the 1000 names obtained, 126 are those of persons born outside the United States. The states which

[1] J. McK. Cattell, A Statistical Study of American Men of Science, *Science*, N. s., vol. xxiv, 2d half of 1906, and vol. xxxii, 2d half of 1910.

[2] The numbers are as follows: Chemistry 175, Physics and Zoölogy 150, Botany and Geology 100, Mathematics 80, Pathology 60, Astronomy and Psychology 50, Physiology 40, Anatomy 25, Anthropology 20.

furnish the most are Massachusetts (134), New York (183), Pennsylvania (66), Ohio (75), Connecticut (40). The percentage in relation to population is four times higher in Massachusetts than in Pennsylvania, and fifty times more than in many southern states. Let us see especially how these men are distributed in the diverse universities. Harvard had 66 in its personnel, Columbia 60, the University of Chicago 39, Cornell 33, Johns Hopkins 30, Berkeley, Cal. 27, Yale 26, Ann Arbor, Mich. 20, Massachusetts Institute of Technology 19, Madison, Wis. 18, The University of Pennsylvania 17, Leland Stanford 16, Princeton 14, etc. About 500 of the names are grouped in 18 establishments; 237 took their fundamental studies at Harvard, 171 at Johns Hopkins, 93 at Yale, 78 at Columbia, 74 at Cornell.[1]

By noting the position of the various names on the list, one can calculate the relative value of the whole of the staff for each science in the various universities or scientific establishments, and Mr. McK. Cattell has arrived at the classification which the table on p. 162 summarizes (in which each number is the place occupied by the corresponding establishment).

Here again, only a very relative importance must be attached to these figures. But on the whole they indicate in which institutions the diverse sciences are, in a general way, best represented. It puts in evidence, likewise in a general manner, the universities which are at the head: Harvard, Columbia, Chicago, Yale, Johns Hopkins, Cornell, and among the state universities,

[1] In weighing these figures we must take account of the fact that certain universities are still very recent. Chicago and Leland Stanford, for example, have only been in existence for twenty-five years.

those of Wisconsin, California, and Michigan. But we must not seek too precise a meaning in each of these figures. Moreover, being based on persons, this meaning

	Mathematics	Physics	Chemistry	Astronomy	Geology	Botany	Zoölogy	Physiology	Anatomy	Pathology	Anthropology	Psychology
Harvard..................	2	1	4	3	3	1	1	1	2	1	3	2
Chicago..................	1	5	10	1	4	4	3	5	10	3	..	5
Columbia................	3	6	7	6	10	..	2	..	7	4	2	1
Yale.....................	4	..	2	5	2	..	6	2	..	..	..	..
Johns Hopkins...........	..	4	5	..	8	10	..	3	1	2	..	..
Cornell..................	7	8	6	..	7	5	5	9	..	..	..	4
Pennsylvania............	..	..	..	9	..	..	..	..	9	7	..	..
Princeton................	6	3	..	..	..	..	10	..	..	..	..	..
Michigan................	..	..	..	10	..	..	..	..	3	6	6	..
Illinois..................	5	..	8	..	..	..	..	..	..	..	..	..
Wisconsin................	8	..	9	8	5	7	..	8	5	10	..	8
California...............	..	..	..	2	..	..	..	7	..	..	5	..
Stanford.................	10	..	..	..	9	6	7	..	..	..	..	9
Clark....................	..	..	..	..	..	..	..	..	..	..	8	3
Mass. Inst. Technology.....	9	7	1	..	..	..	..	..	..	..	..	..
Bureau Standings..........	..	2	..	..	..	..	..	..	..	..	..	..
Department of Agriculture ..	..	10	3	..	..	3	8	..	..	..	..	..
Carnegie Institution........	..	9	..	4	..	9	..	..	..	..	..	..
Smithsonian Institute......	..	..	..	..	6	..	9	..	..	..	1	..
Geological Survey..........	..	..	..	..	1	..	..	..	..	..	..	..
N. Y. Botanical Gardens....	..	..	..	..	..	2	..	..	..	..	..	..
Am. Mus. Natural History..	..	..	..	..	..	..	4	..	..	..	6	..
Rockefeller Institute........	..	..	..	..	..	..	..	4	..	5	..	..
Wistar Institution..........	..	..	..	..	..	..	..	..	4	..	..	..

would be only momentary. Besides, particular conditions intervene for each science; such is the existence of great observatories for astronomy.

As far as regards zoölogy, this table seems to me to represent the reality in the measure in which it can do so, granted that each personality disappears behind the group of all those who compose the department. For this science, and also for general biology, I will add that the American universities are actually in very good condition; that they have produced, of late years, many very remarkable works. Those of Mr. Edmund Wilson, of Columbia, on cytology, for example, are of the first order. From them has come the determination of sex as a function of the chromosomes. Comparative embryology has been the object of extremely precise researches (on cell-lineage), among which must be cited in the front rank those of Mr. E. Conklin of Princeton. We owe to Mr. R. G. Harrison of Yale very remarkable works in experimental embryology, which, in particular, have led to the culture of tissues *in vitro*. The researches of Mr. T. H. Morgan of Columbia, on Mendelian heredity and mutations in Drosophila, are of capital interest at the present hour. Messrs. Calkins of Columbia and Woodruff of Yale, have brought about important progress in the biology of Infusoria, and in the general problems raised by the question of their senescence. At New York, we should mention several other names, like those of Messrs. B. Dean, H. Crampton, and Charles Stockard. At Chicago, the work of Messrs. F. R. Lillie, Tower, Child, Newman, and Patterson; at Harvard, those of Messrs. Mark, Parker, Wheeler, and Castle; at Johns Hopkins those of Mr. Jennings, constitute remarkable contributions in very diverse directions. And one might add many other names to those which I have just mentioned. I do not know whether, at the present time, many other countries could furnish an equivalent.

Physiology, botany,[1] geology, seem to me to lead to

[1] One can judge them in a certain measure from the list of physiologists and botanists forming a part of the National Academy of Sciences. See note p. 224.

similar conclusions. I am not competent enough to formulate a precise conclusion in each of the other sciences.

The development of research in the universities depends on two chief factors: the men who can inspire it and the equipment to carry it out. The first is a necessary condition, the second, though but an aid, nevertheless is very important. Pasteur and Claude Bernard made discoveries which revolutionized biology, under deplorable conditions of equipment, with slight material resources, and almost without co-workers to help them. On the other hand, it is not rare to see sumptuous laboratories from which nothing comes forth, for lack of creative inspiration. But we must not for that discount the importance and value of equipment. If, at the period of their greatest productivity, Pasteur and Claude Bernard had had abundant material resources, as moreover they insistently demanded, their work would have been far from suffering from it, and more than one idea, arrested in the germ state, would without doubt have ripened.

In America at the present time, the equipment is not at fault, and in certain sciences at least, men of ability are not lacking. But it is certain that the material resources have developed much more rapidly than individuals of ability. In a country as rich as the United States, and in which the wealthy classes take an effective interest in the universities, it is easier to build and equip a laboratory than to find a director of the first order for it. Every university aspires to develop as much as possible, and to attract the maximum number of students. For that, it tries to strike the imagination by vast well-equipped buildings which are a tangible argument, for the public.

The luxury and extent of this material equipment, from the confession of many Americans, are often excessive, especially for establishments of second rank. The mark of the spirit of bigness, which impregnates contemporary American thought, is on them.

Harvard does not deserve that reproach. Its present scientific laboratories are, rather, too modest, and call for development and modernization in general, except, however, those of its magnificent Medical School, built a few years ago. The natural history laboratories were still housed, last year,[1] in the Museum of Comparative Zoölogy, founded by Louis Agassiz, and also bearing his name. The finest zoölogical laboratories I have had occasion to visit, are that of Princeton, directed by Mr. E. Conklin, that of Philadelphia (University of Pennsylvania), directed by Mr. McClung, and especially that of Yale, directed by Mr. R. G. Harrison. These various laboratories are less than ten years old. Those of the University of Chicago, also recent, are likewise very well equipped. The laboratory of zoölogy at Philadelphia, very carefully planned by the late lamented Professor Thomas Montgomery, served as a good model for that of Yale (Osborn Memorial Laboratory). There are all the resources desirable for teaching and research, in the various branches of zoölogy (comparative anatomy, cytology, embryology, protistology, physiology). They have not neglected

[1] Those of Botany and Zoölogy are to be moved this year into another building, Pierce Hall, hitherto occupied by the Engineering School, which has been released by moving these courses into the new Massachusetts Institute of Technology in accordance with an agreement between Harvard and the institution.

equipment for keeping living animals (acquariums, vivariums, insectariums, hot-houses). These laboratories have open spaces, into which they can be extended, and which permit experiments in the open air. They have rooms with constant temperature, and equipment for cold. They are constructed so as to be as far as possible free from danger of fire. The cupboards, shelves, book stacks in the libraries, according to a usage growing more and more general in America, are of steel. They are very well arranged as to light and ventilation.

The zoölogical section of the Osborn Memorial Laboratory at Yale cost $1,500,000, not including the instruments. I had the personal satisfaction — though very platonic — of seeing that the plan of this excellent plant answered very exactly the outline of the needs which I had set forth for the new laboratory of Evolution at the Sorbonne, which would be finished today were it not for the war; the credit at my disposal was too modest for me to think of realizing such a program so completely.

Certain universities are well developed for the applications of biology to agriculture, and for the related parts of zoölogy. Such, among others, are Cornell University at Ithaca, N. Y., the University of Illinois at Urbana, and that of California at Berkeley. Harvard, which had, at Forest Hills, near Boston, one of the oldest agricultural schools, Bussey Institution, transformed it, a few years ago, into an institution of experimental biology, devoted especially to the study of Mendelian heredity in animals and plants,[1] and also to

[1] Professors W. E. Castle and E. M. East have made some very important researches there.

forest biology. These various establishments for agricultural biology, the experimental study of heredity, or entomology, as applied to agriculture, are very interesting, and without equivalent in France. I cannot deal at length with it here, and moreover you will find a very precise description of them in the fine book recently published by M. Paul Marchal.[1] after the journey on which he studied the scientific organization of the Bureau of Entomology of the Department of Agriculture (see below).

Without any doubt, remarks of the same kind could be made for the laboratories of sciences other than zoölogy. In all, they have been enormously developed in twenty-five years, and certain of these laboratories have magnificent plants and endowments.

It is especially well known how powerful is the equipment of the American observatories, such as that of the University of Chicago (Yerkes Observatory), and that of California (Lick Observatory).

Besides laboratories, we must not forget the collections and museums, in the equipment of the universities. Harvard has a celebrated Museum of Comparative Zoölogy, founded about 1860 by L. Agassiz, enriched by numerous expeditions, in particular by the oceanographic explorations of Alexander Agassiz. In botany it has the herbarium of Asa Gray (Gray Herbarium), containing today more than 540,000 leaves, and installed, since 1909, at the Botanical Garden, in a specially constructed building, with library (26,000 volumes, and pamphlets), drying rooms, files of photographs, laboratory, workrooms and lecture rooms, all of fireproof materials. Harvard also owns a magnificent col-

[1] P. Marchal, *loc. cit.*

lection of trees, the Arnold Arboretum, near the Bussey Institution, covering more than 225 acres. The museums of mineralogy and geology, of ethnography (Peabody Museum), are no less rich, and these collections are constantly growing, thanks to gifts or bequests of special collections made by professors and specialists.

Finally, of the rapidly growing riches of the American universities the most remarkable is their libraries. This is the number [1] of volumes in the most important libraries in 1913–14 (not including pamphlets).

Harvard	1,200,000	Princeton	320,000
Yale	1,000,000	California	300,000
Columbia	550,000	Illinois	300,000
Cornell	440,000	Leland Stanford	230,000
Chicago	430,000	Wisconsin	207,000
Pennsylvania	421,000	Minnesota	187,000
Ohio	350,000	Johns Hopkins	183,000
Michigan	337,000		

Twelve others exceed 100,000 volumes, and many are those which have between 50,000 and 100,000.

There too, Harvard comes at the head, and its library building, the H. E. Widener Memorial, deserves special mention. It has just been built and was opened for the year 1915–16. The stacks, almost entirely of metal, have a dozen stories and a capacity of about 3,000,000 volumes. Sixty professors have individual rooms in direct contact with the stacks, and can receive their students there. Besides, 300 cubicles, supplied with tables, are arranged in the stacks to permit graduate students who have special authorization, to work near the shelves. It is open from nine o'clock in the morning to ten o'clock in the evening. A printed card

[1] Report of the Commissioner of Education, 1913–14. Cf. p. 162.

catalogue is complete and easy to consult. The librarian has at his disposal more than one hundred employees. On the top floor, thirty-four working rooms, with special libraries of books in current use, are arranged for the work of the students in each of the divisions or departments (Mathematics, French, German, Sanskrit, etc.).

This library includes only the general collection of books of Harvard College, or actually about 675,000 volumes and 433,000 pamphlets. Besides, about 60,000 volumes are scattered in the various laboratories. Finally, about 450,000 volumes more, and 270,000 pamphlets constitute special libraries belonging to diverse parts of the university.[1]

Department	Volumes	Pamphlets
Library of Theology	106,780	50,944
Arnold Arboretum	30,320	7,143
Astronomical Observatory	14,586	34,818
Meteorological Observatory	3,204	15,067
Bussey Institution	3,284	16,067
Dental School	2,228	10,000
Gray Herbarium	15,953	10,672
Law School	161,734	21,989
Medical School	27,000	46,067
Museum of Comparative Zoölogy	52,336	49,219
Peabody Museum of Ethnography	6,328	6,439

I was myself able to test in how practical a fashion this great library is planned, and how liberal and convenient its regulations are. I could also note, for zoölogy and the natural sciences, that duplications are as limited as possible, and how much real wealth the large number of volumes consequently means.

The various preceding notes attest the breadth of equipment of the great American universities, and they

[1] A. C. Potter, *The Library of Harvard University*, 3d ed., 1915.

are particularly impressive when one looks backward
and measures the road covered in thirty years.

The question, then, which is put as to the future, is
to know whether research will become more and more
independent, in what measure it will separate itself
from the college foundation, whether in any case it
really tends to do so, and what will be the resultant of
the diverse influences in play.

A solution for the existing difficulties might be the
formation, in the very bosom of the universities, of well-
equipped laboratories of pure research (the endowment
remaining applicable for research in a large number of
university laboratories is small). If it is essential that
teaching should be in an atmosphere of research, that
does not exclude the existence of certain parts of the
university in which research should reign exclusively.
Foundations of this kind are beginning to develop.
The maritime stations, the observatories, are more or
less in this position. At Harvard the Wolcott Gibbs
Laboratory has recently been created, thanks to gifts,
specially planned for research in physical chemistry,
and directed by Professor T. W. Richards, who was
awarded the Nobel Prize in 1915. Senator Vilas has
bequeathed to the University of Wisconsin, the necessary sums to create ten chairs of pure research, without
routine work, in which the salary of the professors,
which is to be $10,000, would attract men of worth.
There is a very clear tendency to create special institutions of research for each science, of which different
countries possess more or less numerous examples.
The Pasteur Institute of Paris was one of the proto-
types. Germany, in the years which preceded the war,

was systematically creating great institutions of this kind, under the auspices of the Kaiser Wilhelm Gesellschaft.

In my opinion, that is one of the essential forms of the organization which is required. At present, and in the years which preceded the war, the public powers, in France, did not give sufficient attention to it. They had too much superstition for the chair and for oral teaching.[1]

In the United States it seems that the institutions for pure research have won their cause from henceforth.

C. S. Minot, professor of embryology in the Harvard Medical School, who was teaching as exchange professor in the University of Berlin in 1911–12, expressed, in his opening lecture,[2] the idea that America was about to enter extensively on this path, and he divided the history of higher education in his country into three periods; that of the colleges, which is past, that of the universities, which is the present, and that of the special institutions for research, which is beginning. The United States already has, in fact, a rather large number of institutions of this kind, either attached to the universities and more or less autonomous, or completely independent. The most important we shall pass now in review.

[1] Cf. M. Caullery, L'Évolution de notre enseignement supérieur scientifique. *Revue du mois*, vol., iv, 1907.

[2] *Science*, December 6, 1912.

CHAPTER XIV

INSTITUTES OF RESEARCH

1. Research in the service of industry. The Mellon Institute at Pittsburgh.
2. Wistar Institute at Philadelphia. 3. The biological stations: Wood's
Hole, Bermuda, San Diego (Scripps Institution for biological research).

1. MELLON INSTITUTE AT PITTSBURGH RESEARCH IN THE SERVICE OF INDUSTRY

THE *Mellon Institute for Industrial Research*, at Pittsburgh, founded five or six years ago, is of an entirely new type, and deserves to hold our attention quite specially.

It is an establishment for pure research, but attached to a university, that of Pittsburgh, while keeping to itself its own particular board of trustees and constitution and a very great autonomy. It was established, thanks to a gift of $500,000, made in 1913, by the brothers A. W. and R. B. Mellon; of which sum $250,000 was devoted to construction, $60,000 to the purchase of apparatus, $20,000 to the library. The buildings were dedicated in 1915. But the institute had been at work in provisional quarters since 1914.

The plan of this institute is simple and fertile. A manufacturer has a problem to solve, requiring scientific researches for which he has neither laboratories and equipment, nor the necessary men. He turns over to the Mellon Institute a definite sum to have the research in question undertaken by a competent scientist, whom the Institute undertakes to find. The Institute furnishes its laboratories and general equipment. The specialist chosen works under the direction of the Institute; he signs a regular contract; his researches are secret, and their results are the property of the donor of the subvention. With such an organization, manufacturers are spared the general expenses of a permanent scientific equipment, and the greater difficulties of securing a stable personnel of scientists.

The workers thus engaged by the Institute take the name of fellows; each foundation is a fellowship. They are essentially temporary, with regard to the probable length of the researches, one, two, or three years.

The laboratory is equipped for researches in physics and chemistry, especially electro-chemistry and physical chemistry. It has general rooms, equipped with boilers, electric furnaces, and for experiments at a low temperature, and individual rooms for each fellow. Its creation belongs especially to the great movement, largely accentuated since the war, which has in view the development of chemical industry in the United States and freeing it in many of its divisions, from the monopoly *de facto* held by Germany.

From its foundation, manufacturers have understood the value of this institution. The Institute already has a budget of $150,000. In 1914, even before the opening of the permanent building, thirty fellowships were active; some of them require the collaboration of several persons. The subjects bear on problems with which the most varied industries are concerned, smoke-consuming devices, baking, utilization of fruit pulps, hardening fats, high potentials and chemical reactions, turbines, crude petroleum, manufacture of food products, fertilizers, cements, radiators, glassware, natural gas, soaps, metallurgy of brass, yeasts, fertilizers, etc. Naturally the exact nature of the questions studied is not published. The subscriptions amounted to $183,000; and for the current year, the total subscribed for researches was $97,400. A company of ten fellows were engaged on smoke-consuming devices, representing varied specialties, engineers, electricians, a meteorologist, a botanist and a bacteriologist. This fellowship alone was provided for three years with subsidies of $12,500, $15,000 and $12,000.

In a large number of contracts a bonus is promised the fellows, once their research is ended, over and above their regular allowances, which bonus in certain cases reaches $10,000, or a percentage on the industrial exploitation of the process studied. Certain of the fellows have already entered, at the expiration of their researches, into the companies for which they had worked.

This mode of association of science and industry[1] seems to me extremely flexible and practical. It may greatly stimulate young workers; it spares manufacturers enormous general expenses, and permits them to engage in research on a problem, with a clearly limited budget. The facts, moreover, seem to assure, from henceforth, the success of this foundation.

2. THE WISTAR INSTITUTE AT PHILADELPHIA

The Wistar Institute[2] is devoted exclusively to scientific research, and especially to anatomy and embryology. It bears the name of a professor of anatomy in the University of Pennsylvania at the beginning of

[1] Cf. *Science*, May 8, 1914, p. 672, and March 19, 1915, p. 418.

[2] Cf. Bull. no. 5 of the Wistar Institute (Organization and Work of the W. I.).

the nineteenth century, and the generosity of a grand nephew of this scientist, General J. Wistar, made its building possible and assured its endowment.

Wistar Institute is attached to the University of Pennsylvania, which elects annually its board of managers, to the number of nine. Its scientific staff is formed of about ten persons. An advisory board suggests questions it would be desirable to study in the laboratories.

This institution is about twenty years old. It has research laboratories, which have tried to be above all as completely equipped a centre as possible for the study of the brain and neurology.

The institute is at the same time a publishing centre for the journals of morphology: *Anatomical Record, American Journal of Anatomy, Journal of Morphology, Journal of Experimental Zoölogy, Journal of Comparative Neurology.*

Finally, they have tried to make of it a centre of organization for anatomical work (preparation of materials, organization of means of demonstration).

3. THE BIOLOGICAL STATIONS

The Laboratory of Marine Biology at Wood's Hole.

As in other countries, marine biology has been, in the United States, in the last forty years, one of the parts of science in which research has been best coordinated with teaching, and has caused the rise of the most institutes for research. The chief establishment of this kind, in America, is the Wood's Hole laboratory at the southern extremity of Massachusetts, in Vineyard Sound, not far from the tip of Nantucket.

There are, in reality, two distinct biological laboratories at Wood's Hole, one belonging to the Federal bureau of Fisheries, and forming part of the great scheme of governmental scientific services, to which I shall return later; the other independent, which plays a great part in the scientific life of the United States.

Marine biology owed its rise, in America, in large part to Louis and Alexander Agassiz. Under their influence, more or less ephemeral stations were first created. In 1888 they were stabilized at Wood's Hole. At present a certain number of universities and colleges are associated to organize there a well-equipped laboratory. One of the best American zoölogists, Charles O. Whitman, once professor at the University of Chicago has been its soul. This laboratory has been a centre of attraction to which each year the greater part of the most distinguished biologists of the various universities have come to work; a significant example to contrast with the sterilizing dispersion which has been our rule in the matter of zoölogical stations, as in other things.

Thus little by little, Wood's Hole has become a sort of summer capitol of American biology. A large number of zoölogists, physiologists and even botanists bought lands there and each built a cottage. Now, from June to September, there is a veritable congress, more and more numerous, of men like E. B. Wilson, T. H. Morgan, J. Loeb, F. R. Lillie, R. S. Lillie, H. H. Newman, W. Patton, H. V. Crampton, G. N. Calkins, G. Drew, Edward G. Conklin, G. Lefèvre, C. McClung, A. P. Matthew, G. T. Moore, etc., without counting less regular guests.

I saw this city of biologists only in May, when they had not yet come, and unfortunately I could not avail myself of the invitation which they had given me to come and see it during the vacation period.

For a long time Wood's Hole has consisted only of wooden buildings, laboratories and club houses, for the material organization of life is never lacking side by side with the intellectual equipment. Although many universities and colleges were associated to bring it into being and to sustain it, the existence of the Wood's Hole station was rather precarious. But a few years ago, a Chicago benefactor, related to some biologists, Mr. C. R. Crane, came to consolidate the present and even the future. Thanks to his gifts, the research laboratory has been rebuilt of brick and equipped in a complete fashion (aquariums, circulation of water, instruments, library), for experienced investigators; forty to fifty can work there comfortably. At the same time, Mr. Crane has assured the establishment an annual budget which the subsidies of the institutions which send workers there complete. Besides, the station derives about $15,000 from the sale of marine animals to the various universities for the needs of their practical teaching. Thanks to these various resources, which together exceed $30,000, Wood's Hole now has its existence assured, a sufficient equipment, a good fleet, and a permanent staff. Mr. G. Drew, who is resident naturalist, has put it in excellent form. The general direction of the institution is entrusted to Mr. F. R. Lillie, a professor at the University of Chicago.

Wood's Hole has been, and remains, at the same time, a centre of teaching. The old wooden buildings remain and are reserved for the young workers,[1] or for students who come there for instruction. Every year theoretical

[1] A workroom for the season costs $100.

and practical courses are organized in a very precise way for six weeks, from June 15 to the first days of August, on the diverse branches of biology (comparative anatomy, embryology, physiology, general biology, botany). Excursions complement them, and the whole forms a very methodical introduction to the marine fauna and flora.[1]

It will be noted that nothing is free, even in matters of pure science, in America. (University extension and Chautauqua teaching, besides, though works of popular usefulness, pay for themselves.) This system can be criticized, but it has as a consequence that all the tasks undertaken are carried out seriously, or are dropped early. Success is lasting only if the paying participants are satisfied. There are too many things in France which are free, but carried out in too insufficient a fashion, and the general feeling of false shame at accepting or asking for a return for certain exceptional serv-

---

[1] These courses are given by professors or assistants from the various universities. Enrollment costs $50 for each. Here are some figures, relative to the attendance at Wood's Hole during late years.

		1911	1912	1913	1914	1915
Investigators...............		82	93	122	127	137
Experienced	zoölogists..........	42	44	58	50	69
	physiologists.......	18	14	17	22	20
	botanists..........	8	10	11	10	6
Beginners	zoölogists...........	12	21	21	31	36
	physiologists........	2	2	7	1	4
	botanists...........	...	2	7	3	2
Students		65	67	69	89	105
Zoölogists.....................		26	24	33	43	47
Emeryologist...................		20	15	22	21	37
Physiologist..................		6	11	8	10	15
Botanists.....................		13	17	7	15	6
Totals.....................		147	180	191	216	242

ices is certainly bad from the point of view of the collective interest.

Lectures, too, are given at Wood's Hole, and they were the origin of the *Biological Bulletin of the Wood's Hole Marine Laboratory*, today one of the most interesting biological periodicals of the United States, which contains, in preliminary form, a considerable number of works, and of very varied subjects.

Wood's Hole offers us, then, another new example of the realization of great scientific institutions through private initiative and the spirit of coöperation. The property of the station belongs today to the *corporation* of all those who contributed to the foundation, or more than 300 persons, individuals or collectivities; its administration is entrusted to a board, composed of representatives of the various branches of biology. As Mr. F. R. Lillie remarked, in July, 1914, at the dedication of the new building, it answers a very democratic conception: "Freedom of organization," he said, "is one of our mottoes; coöperation is the other. Both are essential and inseparable. In freedom similar interests coöperate naturally, and as long as they respect the freedom. . . . The property and the control of this laboratory are in the hands of those who use it, and that is the essence of a democratic organization."

Other Biological Stations. Bermuda, San Diego

I shall place beside the data on the Wood's Hole station, some concerning the other biological stations, without pretense to completeness.

The Federal Bureau of Fisheries, as has been said above, has a laboratory of its own at Wood's Hole, quite near, but absolutely independent of the one which has just been described. The two stations have, however, often helped one another in various ways. The Fisheries station has been directed especially toward biological research which pertains to the fish industry. The steamer *Albatross*, which is assigned to it, has been placed at the disposal of Alexander Agassiz several times, for his great submarine explorations in the sea of the Antilles and the Pacific.

The Bureau of Fisheries has another station for marine biology at Beaufort, N. C., and is establishing a third at Key West, at the southern point of Florida, to study the subtropical fauna of the Gulf Stream and of the depths which it covers. It is at present planning a fourth on the Pacific. It has, besides, established a fresh water biological station on the Mississippi at Fairport, Iowa. These various stations are open to all qualified scientists.

There is a biological station at the Tortugas Islands in southern Florida. I shall have occasion to speak of it in connection with the Carnegie Institution.

Harvard University has established an interesting biological station in Bermuda, and I had the pleasure of being its guest for a few days. It is modestly installed on an islet, Agar's Island, in unused buildings of the English naval station. The Bermudas offer the naturalist a very rich fauna, extremely interesting for its subtropical character. The lagoons which surround the actual islands are enclosed within coral reefs, which shelter the brilliant fauna usual to these formations. The land is no less interesting than the sea. The few days which I passed at this station are numbered among my best memories as a zoölogist.

The project which should have been realized in the Bermudas was vaster. Several universities were to have coöperated, likewise the local authorities. But this association has not yet been realized.

On the Pacific I had an opportunity to visit the biological station at San Diego, attached to the University of California, and I received there the cordial hospitality of its director, Professor W. E. Ritter. This station is situated a little north of the rapidly growing town of La Jolla, about fifteen miles from San Diego. Its official name is Scripps Institution for Biological research. Again, it is the generosity of benefactors, Mr. and Miss Scripps, which has assured its rapid and considerable development.

Of five planned, two large buildings are already constructed; the first dates from 1909, the second was to be dedicated a few weeks after my visit, and is later destined to be a library. The station is located on a very fine site, on the seashore, on which it abuts for nearly half a mile, and its lands cover one hundred and seventy-five acres. It has a very fine wharf, especially constructed for the use of its fleet.[1]

Its administrative constitution also is impressed with a very great spirit of liberalism. It is attached to the University of California (more than twenty hours by rail from San Diego), which is a state university, but it has as large an autonomy as possible. It is directly managed by its *board of directors*, which includes its director, the permanent members of its scientific staff,

[1] See W. E. Ritter, The Marine Biological Station of San Diego, Its History, Present Condition, Achievements, and Aims. Univ. of California Publications, Zoölogy, 1, ix, no. 4, 1912.

and the donors, Mr. and Miss Scripps. The most important decisions are submitted for ratification to the Board of Regents of the University. Here also, as at Wood's Hole, and contrary to what happens in the universities, the scientific staff has a large place in the effective management of the institution.

The program of this station has been considerably broadened. At first they had in view only the study of marine fauna. Today the general relations between the marine fauna and the flora and the conditions of the environment are the principal object. This includes almost all Oceanography. The station purposes to study the problems of the same order for the terrestrial fauna, thanks to the advantages of the land which it has. Mr. Sumner has undertaken, with this intent, very interesting researches on the variations of Rodents of the American West, belonging to the genus Peromyscus. The director, Mr. W. E. Ritter, is animated with the noblest enthusiasm, and with the desire to contribute, through the scientific work of this station, to general progress. The Scripps Institution is essentially a research laboratory, but as at Wood's Hole, they have organized temporary instruction, in the summer, chiefly for the students of the University of California. Popular lectures are also given.[1]

[1] Several biological stations are now being created in the northwestern states, Oregon and Washington, on the Pacific Coast.

This would also be the place to speak of the botanical gardens, some of which are very large — New York's, at Bronx Park, has more than two hundred and fifty acres; the Missouri Botanical Garden, at St. Louis, has six hundred and twenty-five — and of the great zoölogical gardens. For the last named, see G. Loisel, *Archives Missions Scientif. et Litter.*

The great national parks (there are twenty of them at present), in which nature is rigorously respected, would also furnish admirable biological stations. But they have not yet been utilized in this way.

CHAPTER XV

INSTITUTES FOR RESEARCH

The Carnegie Institution of Washington. Its organization. Its various departments. The Rockefeller Institute for Medical Research at New York.

1. THE CARNEGIE INSTITUTION OF WASHINGTON

MR. ANDREW CARNEGIE has devoted almost all his immense fortune to philanthropic works, and above all, to educational works, the latter being, for him, the fundamental factor in social progress. One of his great foundations is intended to reward acts of heroism; it is well-known and does not come within the plan of this work. We spoke above of the Carnegie Foundation for the Advancement of Teaching.

Finally, the Carnegie Institution of Washington, is intended to facilitate pure scientific research. We will glance summarily at its organization.

Its headquarters are at Washington and it is directed by a president, Mr. R. Woodward, and a board of trustees including twenty-four members; scientists, like Messrs. S. Flexner, C. D. Walcott, Welsh, etc.; business men, financiers, or men prominent in politics, like Mr. Elihu Root. . . . It was founded in 1902; in 1912 it had received from Mr. Carnegie $22,000,000, invested at five per cent. So it has an income of about $1,000,000. Its aim is to encourage research in the broadest and most liberal manner, with a view to the discovery and application of science to the amelioration of human conditions. The means is, "to discover the

exceptionally endowed men in all specialties, whatever
may be their origin, whether they are in the schools or
outside, and to give them the necessary financial aid
in order to permit them to accomplish the work for
which they seem specially designed." Indeed Mr. Carnegie likes to say that his personal successes are due to
the fact that he knew how to find and put at the head of
his enterprises, men better endowed than himself.

The organization is composed of an administrative
division and a publication division, both at Washington,
of a series of laboratories created by the institution at
diverse points, and finally of subsidies given to various
scientists, working in universities or in other establishments.

Actually the Carnegie Institution has created ten
special research departments, listed below, with the
sums which were devoted to them in 1913.

1.	Department of botanical researches.	$37,905
2.	Experimental station of researches on evolution	37,477
3.	Geophysical laboratory	75,000
4.	Marine biological station, Tortugas Islands	18,000
5.	Department of southern hemisphere astrometry	26,316
6.	Nutrition laboratory	48,539
7.	Mount Wilson Solar Observatory, California	254,075
8.	Department of terrestrial magnetism	97,810
9.	Department of economic and sociological sciences	12,500
10.	Department of historical researches	12,500

The total grants to these departments, which were
$649,222 in 1913, were $732,000 in 1914. Let us see
rapidly how each of them is constituted.

Department of botanical researches. — It is composed
of a laboratory of desert biology, established in 1905,
at Tucson, Arizona, and directed by Mr. D. T. Mac-
Dougal. At Tucson, it has over nine hundred acres of

land, and reserves in the mountains besides; it has as
an annex, several experimental stations, situated at
diverse points of the southwestern desert, at altitudes
between sea-level (Carmel, Salton Sea, Cal.), and seven
thousand eight hundred feet (Santa Catalina, Ariz.).
These laboratories have published important studies on
vegetable chemistry, on the relations of plants to water,
the distribution and dissemination of desert plants. It
is certainly one of the most original establishments for
botanical research in existence. It lends itself to researches, not only on plants, but also on animals. Mr.
W. L. Tower of Chicago carried out in part there his
important researches on the Chrysomelidae (Lepti-
molarsa).

The experimental station for researches on evolution is
located at Cold Spring Harbor, Long Island, quite near
New York, and it is directed by Mr. C. B. Davenport,
who received me there with the greatest kindness. It
occupies two acres of land, and includes various zoölogical and botanical laboratories, an insectarium, etc.
Heredity, variation, the determination of sex, and the
various problems of general biology, whether on plants
or animals, are especially studied there. The equipment permits studies on various animals, or vegetable
species. The staff of the station includes, besides the
director, Mr. Davenport, various scientists, such as
Mr. O'Riddle, who is continuing Mr. Whitman's researches on the determination of sex in pigeons, Mr.
Blakeslee, well-known for his works on the sexuality of
the Mucorineae, Mr. Banta, Mr. Goodale, the botanist
Shull, etc.[1]

---

[1] Mr. Davenport also directs an institution independent of the preceding,
and of the Carnegie Institution, although established in the immediate

The geophysical laboratory is located at Washington, and directed by Mr. A. L. Day. Researches of considerable importance for the knowledge of the formation of the earth's crust have come from it. A flood of new light has been thrown on the genesis of the siliceous rocks and the formation of their elements, in particular of the whole series of feldspaths, has been explained by the laws of physical chemistry. What especially characterizes the activity of the laboratory is that it is working on a vast program in a methodical manner, coördinating the endeavors of the scientists who belong to it. It is a fine example of team work. Mr. Day and his collaborators have made, in these last years, some remarkable observations on the lava flows of the volcano Kilauea in the Hawaiian islands, and in particular on the presence of water vapor in these lava flows.

The geophysical laboratory, through the importance of the results which have come from it, and through its equipment, is an establishment today unique in its specialty. It is certainly one of the most convincing examples of the fecundity of the Carnegie foundations, and it shows the returns which can be had from large grants of money in the hands of a skilful director. As may be seen from the figures given, it has considerable resources at its disposal ($75,000).

The Marine Biological Station of the Tortugas Islands is directed by Mr. A. G. Mayer. It is intended for the exploration of marine nature in the tropics, and for the

neighborhood, and devoted to eugenics. He gathers in this institute the most varied statistics on heredity in man, secured chiefly from inquiries and tabulated on filing cards. The latter are then analyzed, taken apart in multiple entries, classified, and put at the disposition of investigators. This institute (*Eugenics Record Office*), spends large sums annually (about $25,000), furnished by gifts, in particular by Mrs. E. H. Harriman.

study of the biology of the coral reefs. It is well-equipped, and supplied with a good fishing fleet. Unfortunately it is rather difficult of access, and physical life there is rather hard. I believe I understood that after the war it would perhaps be moved to one of the greater Antilles. A large number of interesting memoirs have already come from it, due to biologists who have come there to work, and relative either to the fauna of the Tortugas, or to that of the Bahamas, or to that of distant regions, like Torres Strait, to which an expedition had been organized in 1913 by Mr. G. Mayer.

The Department of Southern Hemisphere Astrometry, under the direction of Professor L. Boss, has for its object the exact determination of the stars of the southern sky, and to this end an observatory was established in 1909, on the eastern plateau of the Andes, at San Luis, in the Argentine Republic.

The Solar Observatory on Mt. Wilson, situated at an altitude of four thousand feet, on the mountains which dominate the beautiful city of Pasadena, in southern California, is directed by Mr. G. E. Hale, to whom we owe researches of the highest interest on the solar spots, and the part which magnetic phenomena play in them. These researches have been carried out thanks to the magnificent instruments with which the observatory has been equipped, and which were in large part planned by Mr. Hale himself, and thanks to the addition of a physical laboratory to the observatory proper. Like the Geophysical Laboratory, the Mt. Wilson Observatory is today an establishment possessing technical resources unique in the world, and also, as has been seen, having at its disposal enormous subsidies ($254,000 in 1913).

The *Nutrition Laboratory* was erected in 1907–08, at Boston, in the immediate neighborhood of the Harvard Medical School, which furnishes it heat, cold, compressed air, motive power, electricity, etc. It is directed by Mr. F. G. Benedict. The laboratory is intended for the continuation of researches which this scientist had begun with Atwater, and which have brought one of the most important contributions to the study of animal energetics, that is, the chemical and calorific exchanges of the organism, or, as it is also called, metabolism. Nutrition leads to problems of this order. Whether it is a question of muscular work, respiration, etc., one is finally led to measures of energy; measures of heat thrown off, with the aid of calorimeters; measures of energy furnished by the knowledge of food consumed, work done, etc.

The field of studies includes not only normal nutrition, but its pathological alterations, in states like diabetes, and naturally all the experimental modifications which can be imagined. The great importance of this laboratory is then evident, to which thirteen scientific collaborators are regularly attached, and which has at its disposal, henceforth, a magnificent outfit of calorimeters, various thermometric apparatus, and apparatus for chemical analysis, rooms arranged for the study of metabolism, especially on man, or on certain animals. The possibilities of utilization of this laboratory, for pure science or social applications, are so to speak unlimited, since they cover the whole field of the physiology of nutrition. Besides its regular staff, it welcomes foreign scientists. As has been seen, its endowment is large ($48,000 in 1913).

The *Department of Terrestrial Magnetism* has for a director Mr. L. A. Bauer, aided by fifteen scientific collaborators. Mr. Bauer was formerly director of the United States Coast and Geodetic Survey (see below). The department has studied terrestrial magnetism especially in the oceanic regions. For this purpose, in 1908–09, a ship was built, entirely free from magnetism of its own, and permitting measurements of high precision — the Carnegie, 150 feet long, 568 tons, a sailing vessel with auxiliary engine. This ship has already made long cruises in the Atlantic, Pacific, and Indian Oceans, and besides, has carried out numerous researches on land, in regions still unexplored from this point of view.

Outside of these scientific establishments properly so-called, the Carnegie institution has two other departments, one of *economic and sociological studies*, directed by Mr. H. W. Farnam, and devoted to the diverse questions of political economy of the United States (population and immigration, agriculture, forests, mines, manufactures, transportation, domestic and foreign commerce, banks, labor, industrial organization, social legislation, etc.); the other, *Historical Studies*, busies itself especially with facilitating, directly or by the publication of documents, researches on the history of America.

Besides these special departments, founded wholly new, and entirely sustained by it, the Carnegie Institution distributes important grants to a certain number of scientists working in university laboratories. It has thus made it possible to conduct many important researches. It will be sufficient to mention here those of Professor Richards of Harvard on the atomic weights, of Mr. H. Jones and his students at Johns Hopkins on solutions, of Mr. W. E. Castle of Harvard on Mendelian heredity, etc. The total of the sums devoted to these grants, in 1913, was $200,000.

The Carnegie Institution, then at the end of fifteen years of existence, has already effected a number of important researches, in extremely diverse sciences, and it has succeeded in obtaining a satisfactory return, which depends above all on the skilful choice of men. It will be noted that it has been able to build afresh, with all the technical resources desirable, new laboratories for the study of new questions. That is a great advantage over the methods to which we are generally reduced in Europe, and which consist in using old institutions for new needs. The equipment, and the staff no less, cannot be sufficiently modernized for them. However, certain persons regret that it has bent its endeavors especially to the foundation of permanent establishments, whose budget is very heavy, instead of remaining entirely faithful to the original idea, which was to discover men, and give them, for the time being, the broadest facilities.

It represents actually one of the vastest and most fecund organizations of research. Its annual budget is a little more than $1,000,000.

2. ROCKEFELLER INSTITUTE FOR MEDICAL RESEARCH, OF NEW YORK

Like Mr. Andrew Carnegie, a son of his works, Mr. J. D. Rockefeller, like him, devoted a considerable part of his large fortune to educational works or scientific studies. He is the principal founder of the University of Chicago, which has received $25,000,000 from him.

His name is found among the big donors to the principal universities; thus he contributed to the building of the Harvard Medical School.

One of his principal foundations is the Institute of experimental medicine which bears his name, at New York. One can say that its plan is modeled after that of our Pasteur Institute. It is a collection of research laboratories, centred about the experimental study of infectious diseases, and extending to all parts of Biology which can bring light to bear on them.

The foundation was decided on in 1901. Two hundred thousand dollars were devoted to preliminary studies on similar establishments in Europe; studies which were made by Mr. S. Flexner, now director of the Institute.

The plans were approved in 1904, and the Institute dedicated in 1906. It originally consisted of one building. It has already been expanded by the addition of a hospital, and of a second group of laboratories, whose construction was finished in 1916. It covers a rather large area, on the banks of the East River, around which large open spaces are laid out in all directions. The actual endowment (capital) of the Institute is about $12,500,000, representing an income of about $600,000.[1]

The Institute includes a number of distinct laboratories or departments; pathology, bacteriology, physiological and pathological chemistry, physiology and comparative zoölogy, pharmacology, experimental therapeutics. The staff includes several men of great ability, and numerous works of considerable importance have

[1] The French newspapers announced, in the last days of May, 1917, that Mr. Rockefeller was making a very large donation for the reconstitution of the regions devastated by the war, and that at the same time he was adding $25,000,000 to the endowment of the Rockefeller Institute of New York.

already come from the laboratories. It will suffice to recall the researches of Mr. Flexner and his students on cerebrospinal meningitis and its serotherapy, and on infantile paralysis; Mr. A. Carrel's work on the surgery of blood vessels, and the culture of tissues; that of the physiologist, J. Loeb, on experimental parthenogenesis, and many problems of general physiology.

The Rockefeller Institute has as an annex, a little laboratory for Mr. J. Loeb at Wood's Hole, Mass., adjacent to that which was considered above, and especially, a large laboratory for animal pathology, which was in construction at Princeton, N. J., in 1916, and whose director will be Mr. Theobald Smith, former Harvard professor and well-known for his discoveries on Texas fever, etc.

The Rockefeller Institute is self-governing. The scientists who compose it share, at least largely, in its direction.

A certain number of similar foundations should be mentioned with the Rockefeller Institute, in particular the Chicago Memorial Institute for Infectious Diseases, founded in 1902 and endowed with $2,000,000, the George Crocker Foundation for the study of Cancer, endowed with $1,500,000, the Tuberculosis Institute, founded at Philadelphia in 1903, and others.

CHAPTER XVI

THE NATURAL HISTORY MUSEUMS AND IN PARTICULAR THE AMERICAN MUSEUM OF NATURAL HISTORY OF NEW YORK

THE great museums form another group devoted to research proper, whose modern equipment, endowment in money and in men, make them especially interesting for the European to study. One finds in them the boldness and breadth of conception, and the rapidity of execution, which are characteristic of American enterprises. Also and above all, one finds in them the enthusiasm which the public and especially the wealthy class puts into the development of these establishments. They are successfully adapted to their double rôle, the education of the great public, and scientific progress.

The majority are of relatively recent creation, and their development has been much accelerated of late years, so that there is the following result: [1]

	Number of Museums Founded	Cost of Buildings
1840–49	1	$200,000
1850–59	2	34,000
1860–69	6	1,277,000
1870–79	7	6,030,000
1880–89	5	560,000
1890–99	20	9,866,000
1900–09	21	14,224,000

[1] *Science*, July 26, 1912.

The geographical distribution of the museums in America seems interesting to summarize:

		Cost of Buildings
New England States	19	$4,910,000
Middle Atlantic States	16	17,478,000
North Central States	16	8,466,000
Washington, D. C.	2	4,400,000
Far West (Rocky Mountains and Pacific)	10	1,831,000
Southern States	2	140,000

The National Museum at Washington, maintained out of the federal [1] funds and quite recently reinstalled in a sumptuous edifice, must be placed in the front rank. This museum has been enriched with extreme rapidity. The last report of the Smithsonian Institution shows for 1914–15 the accession of more than 300,000 specimens, two-thirds of them for zoölogy and paleontology. It is evidently destined to become the richest on the continent. Among the other great museums,[2] I shall mention the Carnegie Museum at Pittsburgh, which has a very important paleontological collection, the Museum of the state of New York, at Albany, also rich in this respect, the Field Museum at Chicago, and especially the American Museum of Natural History at New York.

I will restrict myself here to speaking of the last-named, of which I saw more, and which, moreover is the

[1] 1915 budget: $383,500; furniture, $25,000; heating and lighting, $46,000; collections, $300,000; books, $2000; postage, $500; care of buildings, $10,000.

[2] The university museums should be added to this list: some are very large, like the Museum of Comparative Zoölogy at Harvard. Others, less extensive, are nevertheless very rich, for certain divisions. Such is that of Princeton University, which has magnificent wealth for the fossil Mammals, thanks to the activity of Professor W. B. Scott. Such is also that of Yale, which includes Marsh's rich and famous collections, and others.

most accessible to foreigners. It is interesting for its size, its plan, and its working. It is one of the finest institutions of New York.

It is situated quite near Central Park, and its foundation goes back to 1869. Happily, it was conceived on a very large scale: only three-fifths of the plan are actually realized. Its financial working has been based on the collaboration of the municipality and the public. In fact the city has given the land and building, and assured the expenses of the physical upkeep, but it is for private initiative to provide the increase of the collections. It is managed by a board of trustees and its present president is Mr. H. F. Osborn, the well-known paleontologist. As the prosperity of the museum depends in part on the public, its management does everything possible to win its favor. As is logical, the museum is double; there is the museum for popular education, and the scientific museum proper. The first has been planned in the same fashion as the entrance hall of the British Museum of Natural History at London. It does not seek to pile up before the eyes of the overwhelmed and bewildered public, collections of innumerable objects without meaning for it, but to present significant examples in as self-explanatory a form as possible. Whence, for example, the system of groups for the animals, in which they are replaced in their biological environment. They are presented in the setting in which they live in nature. The series of the groups of birds is particularly fine and varied. The flamingoes and their nest-making, leave an indellible impression with whoever has seen them. Another group represents a pool in the New England woods in spring, with the commonest animals which dwell there.

Still another is a vertical section of the beach sand, in which the worms and other types are in place, as one can find them at Wood's Hole. This system is applied to everything. The natural specimens are replaced, when necessary, by glass models, executed with great perfection. Thus the visitor has before his eyes what the naturalist sees in a coral reef, or in any given biological association. If it is a question of giving him an idea of microscopic animals — Protozoa such as the Radiolaria — they still have recourse to drawn glass models, very skilfully executed. It goes without saying that when the specimens lend themselves to exposition and comprehension, models are not substituted for them.

One of the finest and best presented collections is that of the trees of the United States. America far surpasses Europe in the beauty and variety of its forest flora. Europe has only about fifty indigenous kinds of trees. North America has five hundred of them, a certain number of which are giants like the Sequoias and the great Pines (*Pinus lambertiana*, etc.) of the Sierra Nevada forests. A series of specimens marvelously chosen from these kinds, and admirably presented, occupies a large room. It is due to the munificence of Mr. M. K. Jesup. Explanatory documents, tags, photographs, comment on all the specimens in the most educative manner, from the species of the California forests to the Florida mangrove, whose germination one can follow.

Vertebrate paleontology is represented in this museum by admirable material, results of the great explorations of Cope, Osborn and other American paleontologists. The visitor marvels at the admirably restored skeletons of the great secondary reptiles; he reads there without difficulty the evolution of the Equides — from the Eohippus to the horse; of the Titanothera and of many families of Mammals, which is clearly explained to him, always without overloading with specimens among which he would get lost. They show him only just what is necessary in order to understand. The museum of researches and documents, which interests investigators only, consists of stacks and laboratories which occupy the upper stories, where only experts penetrate. Perhaps its area is too limited in relation to the whole. The procedure adopted for exposition, so advantageous from the educational point of view, requires an enormous space.

American mineralogy and ethnography are also magnificently represented.

The action of the museum on the public is not limited to the exposition of specimens and groups. It is complemented by a very methodical organization of lectures, and by the gathering of an enormous collection of projection slides. Rooms and series of slides are put at the disposition of school teachers, who can come to the museum to give series of lectures to their pupils; or of qualified persons for public lectures on various scientific subjects. Circulating collections are lent to the primary schools, in order to show the children significant biological facts. The museum is thus very popular. It receives an enormous number of visitors (1,043,582 in 1909), whom it interests and really instructs. It is open not only during the day, but about one hundred and eighty evenings in a year, for lectures. In 1909 the popular lectures had 82,178 auditors, and the lectures on tuberculosis 42,627. In July, 1916, at the time of my last stay in New York, a congress of museum directors was held at the Museum, under the presidency of Mr. Osborn.

The American Museum is, then, an extremely efficacious instrument of popular education, and it finds, in return, an effective audience among the people. It enrolls as members, by diverse titles, annual members, sustaining members, life members, fellows, patrons, associate benefactors or benefactors, whoever gives from $10 to $50,000 a year. It has, especially, very numerous annual members, to whom it distributes a publication (*American Museum Journal*), which informs them of all the novelties on exhibition. Besides, to measure the aid which individuals bring to it, nothing is clearer than figures. In 1909, of a total expense of $275,419, $160,000 was furnished by the city, and $115,000 by gifts. From 1901 to 1906, $932,000 was spent in explorations and increases of the collections, which came entirely from individual gifts.

The consolidated endowment of the museum has remained relatively small till now. In 1909 it was only a little more than $2,000,000. But it has recently received, through the will of one of its former presidents, Mr. M. K. Jessup, $6,000,000, the income of which is to be used for explorations, scientific researches, and publications.

I have emphasized the popular character. But its true end is the progress of our scientific knowledge, and through its publications and its expeditions of every kind, it has effectively contributed to it. It is above all an institution of research which fully realizes its triple motto: "For the people, for education, for science." In accord with the constitution of the museum, the participation of the public wholly assures its more strictly scientific work. There is no closer combination between a work of public education and of science. In 1910, at a service commemorative of Mr. M. K. Jessup, former president of the museum, and benefactor, Mr. Choate said: "This union of public and private responsibility and generosity which has been the model on which the other similar institutions of the city are founded, has procured for New York something very superior both to the entirely public institutions of foreign cities, and to institutions entirely private in their foundation and management which the other large cities of America possess."

I must add that this museum, with the other great American museums, represents today, in the natural sciences, a very important scientific movement; the publications, bulletins and memoirs, which come from them, and which cannot be enumerated here, fully attest it. For that, they have the indispensable elements at their disposal; first, a large income, permitting them to make purchases of collections on occasion, and especially, to organize excavations, dredging, or expeditions on land, at various points. American activity is very great in this regard, and almost always finds the necessary resources easily. The collections gathered are brought in to enrich the great museums, and to furnish interesting subjects for studies.

A second element, no less indispensable, is an appropriate staff. It must be numerous, for researches of this kind entail extreme specialization, and consequently require numerous curators. In order to have men of ability, they must, on the other hand, be paid suitably. That also leads to a financial question. Nevertheless it is not the only one in question. It is chiefly important for a museum to devote itself first of all to its proper mission, which is, to gather collections, to make them worth while, and to assure their preservation; and not to give courses and lectures. The staff should be selected for their aptitude and taste for doing this kind of research, and not for the qualities which make the professor. Finally, for the presentation of the collections, a museum must have numerous technicians, and real artists, when it is a question of reconstituting fossil skeletons, of presenting living mammals, of making models of microscopic animals, of painting even the reconstitution of extinct animals. That is what the American Museum, among others, possesses at present.

The scientific personnel of the American museums forms a very large body, of unquestionable competency. They meet every year in a special congress, which studies all scientific or professional questions, relative to the organization of museums.

CHAPTER XVII

THE FEDERAL INSTITUTIONS
Scientific Research at Washington

The Smithsonian Institution. The Federal Scientific Establishments: The scientific bureaus of the various ministries (agriculture, commerce, interior). The Geological Survey. Plans for the establishment of a national university at Washington.

The Smithsonian Institution

THIS institution,[1] established in Washington, is the oldest of the great foundations of research in the United States. James Smithson, an Englishman, a member of the Royal Society of London, a friend of Cavendish and Arago died at Genoa in 1829, bequeathing his fortune, in the absence of descendants of certain relatives, to the government of the United States and charging it with the organization of an institution bearing the name of Smithson, "for the increase and diffusion of Knowledge among men." It seems that he got the plan from a phrase in the political testament of Washington, where the great man urges upon his fellow-countrymen the same plan in the same terms. It appears also that he desired to perpetuate his name, in compensation for the disappointments he had undergone because of the irregularity of his birth.[2]

[1] For its history, see *The Smithsonian Institution* — The History of its first half-century, ed. by G. B. Goode, Washington, 1897, 4°, 856 pp.

[2] "The best English blood flows in my veins" he writes; "through my father, I am a Northumberland; and I have royal blood from my mother. But this is of no account to me. My name will live in the memory of men when the titles of the Northumberlands and the Percys will have been extinguished and forgotten." Quoted from *The Smithsonian Institution*, p. 2.

It cannot be denied that the terms of the donation lack precision, and as a consequence, the orientation of the Institute has been and still is somewhat uncertain. Henry, who was its first secretary, had considered giving it the form of a Museum, but believed that this would not correspond with Smithson's intention; nevertheless, under his successors, it is in this direction that it has tended, forming as it did very intimate relations with the National Museum which for a long while was lodged in the buildings of the Institute. Even recently [1] its function has been made the object of discussions. Some would like it to be an institute of research; others reply that a Museum would fulfill such a function. It has been correctly remarked that the absence of precision in the terms used by Smithson allowed great freedom in the administration of the institution and its adaptation to conditions impossible to foresee in the middle of the last century.

Smithson's bequest amounted to five hundred and forty thousand dollars and by means of other donations the capital of the Institute has been increased to a million dollars, almost all of which is deposited at the United States Treasury at an interest of six per cent. In 1915 the Institute spent a hundred thousand dollars in all for its own services (such as $13,569 on publications and $9,021 on special subventions for research work); but it directs on behalf of the government a number of scientific establishments involving a budget of six hundred thousand dollars.

It publishes original memoirs (Contributions to Knowledge and Miscellaneous Collections) and Annual Reports in which it reprints works which it considers

[1] *Science,* first and second half-year, 1906.

worth circulating. All this meets the requirements of the two objects of the Smithson donation: the progress and diffusion of Science. In 1915 the total of its publications made up 6,753 pages with 655 plates, and more than 132,000 copies had been distributed. Moreover, it subsidizes original research and scientific explorations.

The Smithsonian Institute has been the cradle of several federal scientific services which we shall study later, and it is still intimately connected with the National Museum, the Bureau of American Ethnology and the Observatory of Physical Astronomy. To it is entrusted the service of International Exchanges and of American participation in international scientific work such as the *International Catalogue of Scientific Literature*. Its library has been merged with the Congressional Library, of which latter it forms one of the principal parts and represents today more than half a million volumes.

Accordingly the Smithsonian Institute has a rather slender capital in comparison with that of certain of the establishments described above and it disposes of very limited means for the organization of research.

The Federal Scientific Services

The Federal Government, which controls but a small part of the public life of the United States because of the considerable sovereignty of each individual state, has nevertheless been able to develop certain institutions out of all proportion with those of other countries, this being particularly true of the scientific services attached to the various branches of its administration. During the last half-century, it has perceived to an admirable degree the practical value of science and has provided the latter to an increasing extent every year with material means of rendering it useful to the country. It is a question not of science for the sake of science without reference to application, but of the scientific investigation of practical questions.[1]

Washington, the seat of all federal institutions has become through the development of the governmental establishments in question, a considerable scientific centre. There is a *Washington science*, sometimes contrasted with *College science*, the science of the universities, not without a slight flavor of disdain. In reality, both of them reflect, as is natural, professional peculiarities. Their points of view are different. On the one hand, the administrative and on the other the pedagogical atmosphere exert an influence over and manifest themselves among the mediocre element in each of the two systems. The universities are sometimes inclined to multiply their doctors' theses beyond reason in order to demonstrate their vitality; the administrative bureaus on their side tend to seek justification in the eyes of the community for the credits allotted to them in thick reports. But we must not judge them from their defects; the important point is that the faith of the federal government in the practical value of Science and the application of the latter in the governmental services have without doubt helped to increase in a large measure the productivity of the country and to combat the spirit of routine.

[1] It is very significant that all the scientific services have taken rise in the various departments, that there is no department of public instruction in existence. The federal services concerned with instruction of whatever grade constitute a simple Bureau (*Bureau of Education*) under a *Commissioner* and not a Secretary of State. Every state has its own secretary of public instruction or someone equivalent.

The Scientific Bureaus of the Department of Agriculture

This fact is most obvious in the department of Agriculture. Despite the great industrial development of the United States, agriculture has so far been the great source of wealth in the country, and in no country has it made a call upon scientific coöperation to the same degree. The farmers resemble the French peasants very little; especially today the majority of them have received education in colleges of agriculture, even in universities where as we have seen the teaching of the sciences applied to agriculture is given great prominence. Thus they are apt to welcome any information of a scientific nature that may be offered to them.

Now, the Department of Agriculture embraces Bureaus corresponding to the various aspects of agricultural labor; these are veritable administrative establishments of which the total comprises actually more than thirteen thousand government employees and a budget of twenty million dollars. In these bureaus, the scientific services play a very considerable part.

They are as follows, along with their financial provision for 1913–1914:

Weather Bureau		$1,707,610
Bureau of Animal Husbandry		2,031,196
"	Plant Industry	2,667,995
"	Chemistry	1,058,140
"	Soils	334,020
"	Entomology	742,210
"	Biological Survey	170,990
Forest Service		5,399,670

These figures have nothing stereotyped about them; they go on increasing almost regularly with every budget. The endowment of the department of agriculture has passed through the following stages:

Year 1870	$156,440
" 1880	201,000
" 1890	1,669,770
" 1900	3,726,022
" 1910	12,995,274
" 1913	22,894,590

If we consider the Bureau of Entomology by itself, the figures of recent years are not less significant. In 1915–16, when I passed through Washington, the numbers had already risen from $742,000 in 1913–14 to $840,000 and the figure announced for the year 1916–17 is $868,880. Thus, the development of the scientific services continues at a rapid pace. From among all these considerable sums, I invite the reader's attention to the endowment of entomology only as applied in agriculture: $860,000, that is to say about 4,500,000 francs! And yet this figure does not comprise whatever the individual states are spending or the sums spent in the universities and the agricultural colleges.

We cannot enter into even a summary examination of the work of scientific research of these bureaus. Their program of work, published each year (*Programme of Work of the U. S. Department of Agriculture*) constitutes in 1917 a thick volume of five hundred pages; it enumerates, article by article, all the cases of research projected, their object and plan, the laboratories or executive organisms, the names of the responsible persons, the credit assigned, etc.

I will first give certain very brief details on the *Bureau of Entomology*, whose director, Mr. L. O. Howard, very courteously took me round, and I urge the reader to study the institution in greater detail in the well-informed book which Mr. P. Marchal has written about it.[1]

Mr. L. O. Howard has two hundred and five scientific assistants and more than four hundred administrative clerks under his orders. The central bureau at Washington is subdivided into eight sections, each one with its own head, and specializing in the study of insects injurious to a specific class of growths: insects injurious, 1st, to cereals and to fodder farming; 2d, to market farming and to stocks in store; 3d, to fruit trees with caduceus leaves; 4th, to tropical or subtropical farming; 5th, to southern farming; 6th, to forests; 7th, the fight against the Gypsy Moth and the Brown-Tail Moth; 8th, section of apiculture. To each of these sections there correspond a certain number of special laboratories, some quite temporary, others more permanent, ninety-two in all in 1916.

These few facts bring out the importance of this organization. It is conducted with the constant view of rendering really practical service to agriculture. Science comes in chiefly as a factor of economic power; nevertheless, the progress of science as such is thereby greatly furthered, were it only by the amplitude of the information collected and the experiments made in the laboratories. Riley, one of the predecessors of Mr. Howard, succeeded in 1886 in checking the disastrous propagation of cochineal — *Icerya purchasi* — which destroyed the orange trees of California, by introducing an Australian coccinelid — *Novius cardinalis* — which exterminates the cochineal; this method has now become adopted everywhere, and had been applied with success in the region of Nice just before the war. Simi-

---

[1] Marchal, p. xii, *op. cit.*

larly, Mr. Howard has undertaken with success the really gigantic experiment of acclimatizing, in America, the European parasites of the Gypsy Moth (*Liparis dispar*) and of the Brown-Tail Moth (*Liparis chrysorrhea*) in order to check the multiplication of those butterflies which ravaged the trees of New England. This particular form of warfare occupies an entire section of the Bureau of Entomology with a numerous personnel in special experimental stations, and consumes annually more than a hundred thousand dollars.

The need of providing this important institution with an appropriate scientific personnel has created in the United States a considerable school of biological entomology, which has reacted on the universities indirectly and has for example contributed powerfully to the development of biological instruction in Cornell and in the universities of Illinois, California, Nevada, etc.

Let us now consider briefly the *Bureau of Plant Industry*, taking up its budget thus indicating its principal sections; the corresponding credits give an idea of the material importance of the research undertaken by them.

	Number of Scientific Personnel	Budget
1. Central Administration........................	2	$103,880
2. Laboratory of Vegetable Pathology............	10	35,730
3. Collection of Vegetable Pathology.............	4	12,010
4. Research on the diseases of fruits..............	19	69,395
5. Destruction of citrus-canker..................		335,715
6. Research on forest-pathology..................	17	92,421
7. Research on the maladies of cotton, tubercles and fodder plants.............................	14	68,920
8. Research on the physiology and culture of cultivated plants..............................	8	58,840
9. Research on the nutrition of plants............		10,950
10. Research on the fertility of soils...............	20	36,600
11. Research on the acclimation of plants of culture...	13	47,020

	Number of Scientific Personnel	Budget
12. Research on medicinal and poisonous plants and on the fermentation of plants................	19	$65,180
13. Research on agricultural technology............	4	25,220
14. Research on plants with textile fibres..........	2	9,830
15. Research on grains (sampling, manipulation, transport, etc.)............................	42	79,000
16. Research on cereals and their maladies.........	40	140,585
17. Research on maize...........................	12	42,380
18. Research on tobacco........................	15	31,400
19. Research on the plants which yield paper........	3	13,960
20. Research on the resistance of plants to alkalis and drought.................................	5	24,580
21. Research on beet sugar......................	10	42,395
22. Research on economic and systematic botany....	7	34,560
23. Research on dry-land agriculture..............	30	167,120
24. Research on irrigation in the West	12	88,980
25. Research on pomology........................		128,147
26. Research on horticulture and market-gardening..	39	80,333
27. Experimental farm of Arlington		29,880
28. Experimental garden and hot-house, Washington, D. C.....................................	3	54,590
29. Research on the introduction of grains and foreign plants....................................	17	107,080
30. Research on fodder..........................	21	92,980
31. Distribution of seed for experimental purposes....		338,780
32. Practical lectures...........................	10	40,000
	398	$2,488,461

Each of these thirty-one subdivisions is directed by a qualified scientist, among whom we may mention Messrs. W. T. Swingle, W. A. Orton, D. Fairchild, and the total represents, as is obvious, about four hundred scientific workers.[1]

All these items of expense concern research as carried out for the most part in special laboratories or in laboratories of agricultural colleges or in experimental stations. Almost all the experiments are continued for a series

[1] The personnel includes numerous women.

of years. To take an example, since 1904 the bureau has been carrying out methodical experiments in various parts of California on caprification with a view to introducing the cultivation of Smyrna figs. Last year, the results were no less than a production of 6400 tons. To the sole study of the bacteroids of leguminous plants $13,120 were allotted in 1916–17. The investigations upon fruits, upon the acclimatization and the introduction of varieties and new species are of particular interest. But it is impossible to enter into details here.

The Bureau of Animal Husbandry does a similar work of research on domestic animals, their products, their maladies, and especially on all that is related to their economic value. The investigations with respect to the milk industry only entailed, in 1916–17, a credit of $303,270. As an illustration, I may mention such subjects as the study of metabolism in milch-cows, carried out in coöperation with the Pennsylvania State College of Agriculture by Mr. H. P. Armsby, to whom a sum of $3500 had been assigned for that purpose. There is a credit of $452,880 for the fight against the diseases of animals; in 1916–17, a sum of $593,160 was spent for the destruction of the ticks that live on the Bovidae. Scientific researches on the various diseases of cattle are endowed with a sum of $177,160.

The *Bureau of Chemistry* and the *Bureau of Soils* are purely scientific bureaus. Of great interest is the *Bureau of Biological Survey* which studies all the problems concerning the Mammals and wild birds, is occupied with the protection of game and with the care of the territorial reservations for the big animals such as the bison, and studies the behaviour of animals whether indigenous or immigrant, the distribution of the various species, and the migrations of birds ($3,750). Its endowment was $614,530 for 1916–17.

Bureaus Attached to the Department of Commerce

Three great scientific agencies are attached to this department: the *Bureau of Fisheries*, the *Bureau of Standards* and the *U. S. Coast and Geodetic Survey*.

1. *Bureau of Fisheries*. — Among the functions of the Bureau of Fisheries is the economic study of both sea and fresh water throughout the United States and Alaska; and one of its principal aims is to increase their yield by the application of scientific methods to all the biological questions concerning aquatic animals useful to man. Its program is therefore very varied: seafishery; study of the development and habits of sea fish; the breeding of fish in the sea or in fresh waters; the study under the same conditions of the edible molluscs and crustaceans; the stocking of fresh water; not to mention special questions such as all that concerns the fur seals of the Pribiloff Isles in Behring's Sea.

The budget of this bureau was $1,132,390 in 1912; $944,790 in 1913; $1,047,180 in 1914. These sums include about $400,000 for salaries and wages, $335,000 for the propagation of edible fish, and $60,000 for the maintenance of the fishing boats.

As was mentioned above, the Bureau is provided with two maritime stations for scientific studies, one at Wood's Hole, Mass., and the other at Beaufort, S. C.; it is now installing a third to the south of Florida, at Key West, and is planning another for the Pacific. It also has another large station for the biological study of fresh water, on the Mississippi, at Fairport, Iowa. Moreover, there are several stations for fish breeding of a practical nature, and similar operations. For all work at sea, there is a large steamboat, the *Albatross*, built for distant and prolonged voyages and capable of undertaking the deepest kind of dredging operations; a motorboat, the *Fish Hawk*, and other less considerable boats.

The Bureau of Fisheries has frequently organized oceanographical expeditions on a great scale on the *Albatross* and it has been led to a profound study of various biological questions. Thus it was that recently a mission organized under its auspices and consisting of Messrs. G. H. Parker, W. H. Osgood, and E. A. Peeble visited the Pribiloff Islands in the summer of 1914 in order to study biological problems on the spot and collect statistics on the herds of fur seals (*Callorhinus alascanus*). The story of this expedition is most interesting and was published in the *Bulletin of the Bureau* which forms every year a volume of different biological monographs.

2. *The National Bureau of Standards.* — This Bureau, established in 1901 and directed by Mr. S. W. Stratton, has a rôle analogous to that of the *National Physical Laboratory* in England, to the *Technische* and *Physikalische Reichsanstalt* of Berlin and to what ought to be the *Experimental Laboratory of the Conservatory of Arts and Crafts in Paris*. It keeps the standard measures, and makes all the measurements and tests of instruments. It controls the fundamental measures of length, time, and mass, and the electric measures. It also undertakes on behalf of the government measurements of quality, to this end incessantly perfecting the methods of

measurement and weighing or inventing new ones. It would be impossible to give appropriate examples without entering into long technical details.

This last category of measurements is undertaken only for the government. In order to avoid putting obstacles in the way of private initiative, non-official experimental laboratories have been left entirely free.

The Bureau of Standards aims further to determine the value of physical constants that are needed in industry.

Its personnel is distributed according to the nature of the scientific work (general measurement, electricity, heat and thermometry, optics, chemistry, metallurgy) and in 1915 consisted of two hundred thirty-three members of whom a hundred and forty-five were scientific workers.

The building in which it is housed cost one million dollars in construction expenses and half a million for the furnishing. The total budget amounted to $543,645 in 1913, to $637,015 in 1914, and to $695,811 in 1915.[1]

In 1915 the Bureau had carried out 116,204 tests and printed one hundred and thirty-seven publications of which forty-six were new, comprising twenty-six scientific and technical memoirs. By the very nature of the practical services which it is destined to render, this bureau is led to carry on important scientific enquiries in the different branches of Physics.

3. The *United States Coast and Geodetic Survey*, attached to the department of Commerce since 1903, dates from 1807. It is a great geodetic and hydrographic institution concerned principally with triangulation, astronomical measurement, the study of terrestrial

---

[1] Of these $293,500 for running expenses and salaries.

magnetism, topography, the study of tides and tidal currents, the survey of the country, the study of gravity and geodesy in general. It is thus an agency contributing to the progress of science. It used to be a branch of the *Geological Survey* but is now separated from it.

DEPARTMENT OF THE INTERIOR

The United States Geological Survey

The *Geological Survey* attached to the Department of the Interior is a scientific institution of considerable importance and in its actual form, unified as it is for the whole of the United States, it dates from 1879. Before this time, there existed analogous organizations limited to particular portions of American territory. These partial *Surveys* have accomplished a remarkable work in geology and geography. The exploration of the Grand Canyon of Colorado by Major Powell is a classical example. The *Coast and Geodetic Survey*, of which mention was made above was only recently detached from the Survey in question.

The function of the *Geological Survey* is to secure an inventory and a classification of the lands, waters, and various mineral products of the national soil.

To give an idea of its recent development and present resources we will mention that in 1879 it was endowed with $106,000, in 1889 with $801,240, in 1903–04 with $1,377,820, and in 1914–15 with $1,620,520. During the last year, its personnel consisted of nine hundred and nine workers.

Its program, which is at once scientific and practical, may be discovered from its subdivisions: the survey of Alaska, mines and metallurgical resources, chemical and physical research, topography, geography, and forests, hydrography (study of rivers), hydrology (subterranean waters), utilization of surface waters (hydro-economics), publications.

The publications of the *Geological Survey* are very great in number; already six hundred bulletins have appeared many of them with more than a hundred pages respectively, and also sixty large geological monographs, without counting the very numerous publications on surface waters. Finally, this Survey is entrusted with the execution of the general geological map of the United States, an enormous enterprise and still far from accomplishment. During a single year, in 1914–15, the Survey edited sixty-six publications, with 21,407 pages and 191 maps.

The *Geological Survey*, through its explorations and publications may be said to have become an important instrument in the study of pure geology. And the facts which it has been able to collect have been a factor of prime importance in the economic development of the United States.

However specialized its field, it nevertheless contributes to the general education of the public. To give an idea of the dissemination of scientific knowledge effected by these great organizations, I will cite a case from my personal experience. While traveling from Chicago to San Diego on the Santa Fé Railway, on a visit to the Grand Canyon of Arizona, then going back from San Diego to San Francisco by the coast line, and finally returning from San Francisco toward New York by Ogden, the Great Salt Lake and Yellowstone Park, I was able to study and appreciate the scenery along all these lines, thanks to the recent publications of the

Geological Survey,[1] which inform the traveler, for every portion of the line, of everything that is worth seeing from the train, from the point of view of physical geography, geology, the natural resources of the country, and the recent history of its colonization. These publications prove convincingly how science always tries to justify its usefulness by direct and tangible services rendered to all the members of the community.

The preceding list of scientific governmental establishments is not complete. One should add the National Museum, of which mention was made elsewhere, the Bureau of Mines, the Public Health Service, the Bureau of Education, the Naval Observatory, and finally the Library of Congress which is equivalent to our National Library. This library, equipped luxuriously and yet in very practical fashion, offers considerable resources for scientific work. In 1912 it possessed more than two million volumes, and had a budget of about $600,000 of which $100,000 were for purchases.

Washington is unique among American cities, and something of a paradox to Americans, for it has neither commerce nor industry to justify its growth; it is simply a city of administration. Moreover, it is one of the most beautiful cities in the United States built on a very original plan, the work of a Frenchman, L'Enfant, a major of the engineers, who had come to America with Lafayette and Rochambeau. Nowadays, it has an increasing number of marble palaces and its monuments are rapidly multiplying.

As we have just seen, Washington has also become a

[1] Bull. 612. — *The Overland Route* (244 pp.); 613. — *The Santa Fé Route* (194 pp); 614. — *The Shasta Route and Coast-line* (142 pp); with maps — at a scale of 1/500,000 — of all the lines and many photographs, 1915.

considerable scientific centre in that its governmental institutions embrace scientific organizations. From among the thousand best men of science in the list prepared in 1906 by Mr. J. McK. Cattell, one hundred and nineteen resided in Washington.

These considerations have given birth among many minds to the idea of establishing in Washington, rich as it is with so many resources, a great National University.[1] The plans vary as to detail but are in general agreed on the point that such a university ought to be of a different type from those already in existence. It should altogether dispense with elementary education and devote itself uniquely to scientific research. In any case, instruction should be reduced to a minimum. It should likewise dispense with all awards of degrees and diplomas. Above all, it should constitute a better utilization of the immense scientific resources of the federal capital which at the present are being somewhat stifled by the too great administrative atmosphere of the place. The National University, says J. McK. Cattell, would be the best instrument for uniting the ideals of the democracy into a single body.

This project was formulated chiefly by the presidents of the state universities, who see in it a natural extension of the conception on which their own universities are based. The federal government would thus be able to produce, in the way of a university, something beyond the forces both of the individual states, owing to its immense resources, and of the private universities whatever their wealth and the devotion of their alumni, or whatever the resources of a Carnegie or a

[1] Cf. *Science*, August 16, 1912, November 29, 1912, January 17, 1913, February 15, 1914.

Rockefeller. On the other hand, the private universities have shown themselves quite hostile to the plan, seeing in it, with more or less clearness, a threat against themselves and in any case an aggravation of the competition against them by the State. Disregarding the egotistical element in this opposition, it would none the less remain true that a university too powerfully concentrated in Washington would have to balance its advantages against serious disadvantages. One of the favoring circumstances in the scientific evolution of the United States has been precisely the fact that intellectual activity has not been concentrated at a single point, nor in the hands of the State, and that powerful and completely autonomous organizations still seem capable of balancing one another.

CHAPTER XVIII

ACADEMIES AND SCIENTIFIC SOCIETIES

The American Philosophical Society. The American Academy of Arts and Sciences. The National Academy of Sciences: its rôle, composition, and mode of elections; reflections and comparisons. The American Association for the Advancement of Science.

NOWADAYS the United States numbers many academies and scientific societies whose rôle hardly differs from that of the analogous organizations in Europe, except perhaps in that the immensity of the territory provides more of a *raison d'être* for academies or local societies and endows the large national societies with greater importance as a means of coördinating scientific activity.

Let us first consider the oldest of these.

The dean of the large scientific societies of the United States, which even today is among those that have most prestige, is the American Philosophical Society established in Philadelphia. According to its seal, it was founded in 1727 and its complete title is The American Philosophical Society, held at Philadelphia for promoting useful Knowledge. Benjamin Franklin was its founder and first secretary. The title which it adopted under his inspiration reflects the purpose of making science useful to man. The original program enumerated a long series of possible undertakings, including "all experiments of a philosophical character which might illuminate the nature of things, or tend to increase the power of man over matter, or multiply the goods and pleasures of life." The word "philosophy" is here used in its classical English sense so as to mean the total of what we call science.

The society was organized after the model of the London Royal Society. It has been publishing Transactions since 1799, and Proceedings since 1838. It has a national as well as a local character. The general meeting held annually at Easter in Philadelphia is attended by a large number of members from various parts of the United States. I was honored with an invitation in 1916. In accordance with American customs, besides communications presented by the members, the order of the day includes a question which a group of members have been requested to study in advance from particular points of view, giving to the meeting the character of what the Americans call a symposium. In 1916 the topic was the organization of peace.

The session comes to an end with a very cordial banquet of which the menu was embroidered with a sprightly humor borrowed from the best authors. "I believe in banquets, they lubricate matters"; ran a quotation from Lord Stowell. Every course becomes an object of a more or less classical allusion and every toast is announced in similar fashion. The toastmaster invoked with a verse from Troilus and Cressida the privilege of choosing his own subject. From among the other toasts and in accordance with tradition, the first concerned Benjamin Franklin whose memory is particularly vivid in the Eastern United States; in 1916, Professor Trowbridge of Princeton, the orator of the day, evoked before his audience the life of the founder of the society and the decisive part which he had taken in the military organization of the American colonies in the eighteenth century — a procedure analogous to that which was urged in 1916 by the partisans of preparedness. The other toasts concerned sister-scientific societies, universities, and the society itself. Last year, in the thoughts of almost all there was the thought of the European war and sympathy for the cause of France.

The oldest of the American Academies after the Philosophical Society is the American Academy of Arts and Sciences, founded in Boston in 1780, on a model more similar to the Paris academies. It meets once a month, between October and May, in a very comfortable home which it owes to an important bequest from Alexander Agassiz.

This society has more of a local character than the other, although the rather numerous list of its members covers the various parts of the United States. The maximum number of its national members is six hundred, divided into three classes (mathematics and physical sciences, natural sciences and physiology, moral and political sciences).

The Connecticut Academy at New Haven, founded in connection with Yale dates from 1797. Its Transactions dating from 1866, contain the celebrated memoirs of J. Willard Gibbs. The Maryland Academy at Baltimore was founded in 1809, the New York Academy of Sciences in 1817, then designated as a lyceum. Nowadays all the large cities have their own more or less recent academies. The Washington Academy of Sciences instituted in 1898, merits special mention because it forms a federation of sixteen specialized scientific societies established at the federal capital and continuing in independent existence with their own individual publications. The former must not be confused with the National Academy of Sciences, which we shall consider presently.

THE NATIONAL ACADEMY OF SCIENCES

This Academy is equivalent to our own *Académie des Sciences*, or to the London Royal Society, and we shall consider it at some length.

It is recent, having been established by an act of Congress on March 3, 1863, during the War of Secession; thus it is barely more than half a century old. Its act of establishment gives it an official character, though rather vaguely so. In the minds of its founders it was destined to serve as the scientific council of the government, furnishing it with reports on such questions as were put to it. Up to a very recent period, this function had remained only virtual, as it was being fulfilled by the scientific bureaus of the various departments of state. The present war seems about to change this situation. In fact, at its session of April, 1916, the Academy unanimously resolved to offer its services to the President of the United States in the interest of national preparedness, and Mr. Wilson accepted the offer. The plan of the Academy is to coördinate the scientific resources of the various institutions of education and research, utilizing them for the prosperity and security of the nation. The above-mentioned resolution resulted in the creation of a National Research Council, which has already extended its mission beyond purely military problems so as to cover all kinds of industrial investigations or research in pure science.[1]

[1] The Council was composed of scientists and expert engineers not only from the Academy but from the most various institutions. It formed a central committee at Washington under the presidency of Mr. G. E. Hale, the

However, until then the Academy had existed much more as a private society than as an institution of the state. It has received no subsidy, so to speak, and is supported by the membership fees. It has no offices of its own, and it avails itself of the hospitality of the *National Museum* for its meetings at Washington. During the last few years, the construction of a mansion of its own has often been referred to as one of its most imperative requirements, but curiously enough, in this country of rapid realization and rich and numerous endowments, the desire is yet far from accomplishment, although the local academies are often sumptuously installed. Federal institutions hardly interest individuals and Congress does not seem to have much affection for pure science. The Academy disposes of a few rather modest endowments for the carrying on of research work. Alexander Agassiz, whose generosity is evident in many circumstances, bequeathed to it some years ago fifty thousand dollars to use as it pleases. Thus it is seen that generally speaking the Academy holds a very modest position from a material point of view in relation to a number of institutions of infinitely less importance.

Its composition has been modified several times since its foundation, when the number of its members was fixed at fifty; in 1870 it was raised to a hundred and fifty and every year ten new members were elected until

well-known astronomer, and also local committees. In attaching to itself workers from outside its membership on such a broad basis and in a spirit of complete equality, it has given a very beautiful example of truly liberal and scientific spirit.

Considerable sums have been already placed at its disposal by private initiative; $100,000 by the *Troop College of Technology* of Pasadena, and $500,000 by the *Massachusetts Institute of Technology*.

the maximum was reached. By a new modification voted in 1915, the maximum of members has been raised to two hundred and fifty and the number of annual elections to fifteen.

In 1916 there were about a hundred and fifty members distributed among the following nine sections:

		Members in 1916
1.	Mathematics	11
2.	Astronomy	11
3.	Physics and Engineering Sciences	26
4.	Chemistry	25
5.	Geology and Paleontology	26
6.	Botany	10 [1]
7.	Zoölogy and animal morphology	20 [2]
8.	Physiology and pathology	17 [3]
9.	Anthropology and psychology	10 [4]

The Academy members are scattered throughout the United States; one recently published list shows that:

Eighteen members belonged to the federal scientific establishments at Washington.

Twenty-three members belonged to Harvard University, Cambridge, Mass.

Fifteen members belonged to Yale University, New Haven.

[1] Messrs. N. L. Britton, D. H. Campbell, J. M. Coulter, W. S. Farlow, G. L. Goodale, C. S. Sargent, E. F. Smith, R. Thaxter, W. Trelease.

[2] Messrs. J. A. Allen, W. E. Castle, E. G. Conklin, W. H. Dale, C. B. Davenport, H. H. Donaldson, R. G. Harrison, H. S. Jennings, F. R. Lillie, F. P. Mall, E. L. Mark, C. H. Merriam, T. H. Morgan, E. S. Morse, H. F. Osborn, G. H. Parker, A. E. Verrill, C. D. Walcott, W. M. Wheeler, and E. B. Wilson. Messrs. L. O. Howard and R. Pearl were elected to this section in 1916.

[3] J. J. Abel, F. G. Benedict, W. B. Cannon, R. H. Chittenden, W. T. Councilman, S. Flexner, W. H. Howell, J. Loeb, G. Lusk, F. P. Mall, S. J. Meltzer, L. B. Mendel, T. M. Prudden, Thomas Smith, V. C. Vaughan, W. H. Welch, H. C. Wood.

[4] The total gives 156 members. But a good many of them are counted twice over as they belong to two sections at the same time (paleontologists, for example, to those of geology and zoölogy).

Thirteen members belonged to Chicago University.

Eleven members belonged to Columbia University, New York.

Ten members belonged to Johns Hopkins University, Baltimore.

Five members belonged to the laboratories of the Carnegie Institution.

Four members belonged to California University.

Three members each belonged to the following universities: Wisconsin, Madison; Cornell, Ithaca, N. Y.; L. Stanford, California; Clark, Worcester, Mass.; and to the Rockefeller Institute, New York.

Two members each belonged to the following universities: Princeton, Pennsylvania, Michigan, Northwestern (Evanston, Ill.) and the Massachusetts Institute of Technology.

Owing to the fact that its members are so scattered, the Academy cannot hold meetings at frequent intervals. It meets regularly twice a year, once at Easter at Washington, and once toward the end of November in a locality which changes every year. Beside business matters, the meetings are devoted to individual communications from members and to symposia on specific questions.

The Academy had honored me by an invitation to attend its session of Easter, 1916, and there I had the pleasure of meeting, as also at the American Philosophical Society, a large number of the most eminent scientific men in America. Seventy-two members, that is to say, about half the membership of the Academy, attended the meeting; some had come from California, having crossed the entire continent in order to be present at the meeting. The symposium had been organized for that session by Professor W. M. Davis of Harvard on the Methodical Exploration of the Pacific. A series of specialists expounded the plan of research which should be organized in the fields of the various sciences in view of setting up a program and collecting funds later. The Academy is preparing an extensive and long undertaking in order to study the ocean and the territories emerging from it, and to consider problems of the physical and natural sciences, from that of gravity to those of the nature of fauna and flora and questions of ethnography.

The elections are carried out according to a system analogous to that of the London Royal Society. A person cannot become a candidate unless his name has already been proposed by the majority of a section or of the Council of the Academy; this allows discussion only on men whose value is recognized by the experts, and the sections are usually of a sufficiently large membership to prevent exclusion of anyone from personal animosity. It is important to notice that it is specialists who designate members first, such designation being in fact the only rôle of the sections.

At the annual Easter session, all the names which have been printed on the list and voted upon by the sections are submitted to a first vote by the members of the Academy, a vote which cannot include more than fifteen names. The results of this preliminary vote are classed according to the votes obtained, and thus a list of preference is secured.

Every name in this list is then finally voted on, separately and in the order on the list of preference, and the candidate is declared elected when he receives two thirds of the votes given with twenty-five as a minimum. The order on the list of preference is followed until fifteen have been elected or until the total number of members reaches the figure of two hundred and fifty.

No method of election could change human nature or suppress intrigues, but the one just mentioned renders intrigues as difficult as possible, instead of putting a premium upon them — as is the case with our own Academy — by avoiding the method of direct application for candidature. It is infinitely more difficult to take a group of specialists by surprise than the membership of an assembly dominated by incompetents; it is true that often specialists have the defect of one-sided and very exclusive views and that, like all men, they may be partial. But as the greater part of the sections include twenty members respectively, the partiality of two or three either in favor or against a given candidature has no serious chance of causing the appearance or the removal of a name on the first list of candidates.

To my mind, then, it is a great advantage that the membership of the Academy should be so numerous. Mr. G. E. Hale, its very eminent foreign secretary, has devoted an extremely interesting article to the subject of the rôle of Academies in *Science*,[1] and has made a comparative study of the academies of large countries. He brings powerful reasons for not making academies into very closed bodies and chooses in favor of the system of the London Royal Society — the English equivalent of our "*Académie des Sciences*" — which has four hundred and eighty members at the present time.

In Europe, "on the continent," he says, "I have known of scientists who did not form part of academies and did not receive the aid of neighboring universities, of men who could not be elected into the academies because the number of members of the latter was too limited or their traditions unchanging. In England, such men would have been admitted into the Royal

[1] *Science*, November 14, 1913, February 6 and December 25, 1914, January 1, 1915.

Society as fellows, and the Society would have been happy to publish their memoirs or to aid them in some other manner. . . . By taking in a larger proportion of young men actively engaged in research, the Academy has increased its contact with living issues, and thus made itself more truly representative of American Science. . . . The purpose of an academy" — adds Mr. Hale — "is not merely to confer distinction by election to membership, but to constitute a working body."

The National Academy of Sciences aims to include in its membership all American scientists of distinction. It has an indisputable moral authority in the United States, but exercises no effective power. Accordingly, it does not obstruct in the least, the growth of the various scientific institutions, the universities or the establishments which we have reviewed and each of which has an independent existence.

As can be seen, the constitution of the National Academy of Sciences differs greatly from that of our Academy of Sciences, and to me it seems better adapted to present conditions. Our Academy bears the weight of a past which has been glorious but which chains it the more so that, in contrast with its American sister-academy, it is not free in its movements. It is joined to the other sections of the Institute, and the latter not being as a whole founded to serve the scientific spirit, is more given to conservatism than to audacious reforms.[1] Whereas so many things have been renovated

---

[1] A witticism of Monsieur Paul Bourget's, of some years ago, has been recently revived (Leon Bloy, *Au seuil de l'Apocalypse*, p. 36 and P. S., *Le Temps*, August 11, 1916) to which the war has given a particularly piquant relief, and which expresses, as it were, the paroxysm of the state of mind in question: "Four barriers," wrote Monsieur Bourget, "separate us from barbarism: the great German general staff, the English House of Lords, the Institute of France, and the Vatican."

during the century, the Institute still keeps, without any retouch so to speak, the statutes granted to it by Bonaparte, with the costume designed for the pompous ceremonies of the Consulate. The Academy of Sciences retains its eleven sections with six members each, established in accordance with the state of knowledge at the end of the eighteenth century, but of which the numerical equality and limitations are no longer in harmony with the relations among the sciences at present.

Up to about five or six years ago, it was necessary that one should live within the circle of the Paris fortifications in order to be a member at all, and this simple detail in the regulations inspired by a period when there were no trains in existence has had disastrous consequences for the vitality of science in the provinces. The Academy abolished this restriction, though tardily and not without timidity and reservations. The discussion in which the whole Institute took part has shown the excessive importance, in the eyes of a good number of its members, of the possibility of depreciating the value of the title which they hold by the multiplication of the number of its holders — a preoccupation which recalls the Duke of Saint-Simon rather than modern society.

Under its present constitution, the Academy receives almost all its members too late, the larger part after the really productive portion of their career has passed. Thus, its influence which is in fact very great, is exercised by men the majority of whom have already passed the age of enterprise and the outlook toward the future. Unavoidably, a community dominated by aged men has the tendency to distrust whatever seems to upset the conceptions to which it is accustomed. Particularly among the sciences, if one looks back a century, how many successive transformations and cases of a rise and fall of theories, that seemed to explain everything, will one not observe — of theories that have had their moment of great fertility but that have yielded their place to others! The examples would be numerous in Chemistry, Physics, and Biology. It may be the atomic theory, the optics of Fresnel, and later that of Maxwell and the theory of the electrons; it may be Darwinism. The conceptions of tomorrow do not destroy those of yesterday; but in order to contemplate the relations of phenomena in a new light one must be capable of sufficient self-detachment. Doubtless, the mind of the scientist is aware of the essentially transitory and relative rôle of the hypothesis. Yet despite everything, one becomes attached to the hypothesis one has employed during one's maturity, and one becomes more or less incapable of thinking without it, especially of foreseeing the fortunes and fertility of those theories that are to succeed it. To mention but one example only, the Academy had for long taken a hostile attitude to the Darwinian movement and its section of Zoölogy refused to receive Darwin himself.

It is not desirable that the scientific body, endowed as it is with the greatest moral authority, be composed chiefly of men who are at the end of their careers. Such a condition leads inevitably to a gerontocracy tending to inhibit the *élan* of the younger generations. It is necessary that the latter should have all the possible means of action at their disposal; even so, they encounter not a few obstacles to progress. Without at all suggesting that scientists whose whole career has demonstrated their excellence should be excluded, we urge that these should be associated in a greater measure with young men, and that the latter should be placed in a lesser degree under the tutelage of their elders, or more exactly of six of their elders in each science, who are never, all of them, great men and who owing to their small number are possessed of excessive power.

Owing to the fact that access to the Academy is so restricted and that it depends to such a degree upon circumstances, the title of a member of the Institute becomes above all a personal distinction, the consecration of a career, a sort of superior decoration, the prestige of which in the eyes of the public is doubtless the sign of a certain idealism, but the very difficult conquest of it has the effect of restricting the freedom in mental direction of more than one scientist. There are thus several circumstances to which the Academy is indifferent in appearance but to whose evil effect its influence contributes, in a more or less decisive fashion, respecting the grading and the selection of members, that is to say, in the last analysis, respecting scientific production.

The function of the Academy nowadays, since science is cultivated outside it and since many specialized scientific societies are in existence, is above all that of coördinating the various sciences together, and this would be better accomplished and in a more wholesome fashion, were the gates of the Academy more widely open.

It appears to me then that the Academy of Sciences, following the example of the Royal Society and the American Academy, should enlarge itself, redistribute its sections, determine for them neither equality of numbers nor any numerical limitation, but merely establish a high maximum of membership, a maximum which generally should not be attained. Vacancies would thus always exist and the regular access to membership be possible to men of value as soon as their excellence is duly ascertained and at the period of their full activity in research. So far as this point is concerned, a method of election such as the one in vogue at the Royal Society or at the American Academy would regularize the automatic renewal and invigoration of the institution; it would moreover have the advantage of suppressing the direct application for candidature; for it is much more natural that a scientific body should discover of its own accord the men whom it would like to have as members.

THE AMERICAN ASSOCIATION FOR THE ADVANCEMENT OF SCIENCE, AND SPECIALIZED SCIENTIFIC SOCIETIES

Nowadays, scientific societies in the United States are extremely numerous and more and more specialized; to enumerate them would be out of the question; some are local,[1] others are national, but owing to the immense extent of the country, the latter tend to create local sections in the large cities.

I will say a few words about the rôle of the American Association for the Advancement of Science, which is equivalent to our own French Association. It was founded in 1848, on the model of the British Associa-

---

[1] Among the oldest, I will simply mention, so far as biology is concerned, the *Academy of Natural Science* of Philadelphia which celebrated its centenary in 1912; it has in important library and its publications are considerable. The *Boston Society of Natural History* dates from 1830.

tion; [1] it has two classes of members: professional scientists or fellows, and others who are simply interested in science, or members. It is specially interested in serving as a connecting link between many societies that are specialized. It meets twice a year at different cities, once in August and once in the last week of December. The latter meeting is the more important in that a very large number of societies meet during that convocation week and the majority of them in the same locality as the American Association itself. Thus, in December, 1915, eighteen societies met simultaneously at Columbus, Ohio. This is indeed a very fortunate habit, and the congresses of the French Association would gain a great deal if by an understanding with the special societies the annual reunion of the former could be made to coincide with the meetings of the latter.

Every year, the American Association organizes symposia in its various sections, by means of which important questions may be examined from varied points of view.

A European is a little confused by the multiplicity of the meetings which an American scientist is invited to attend every year, in very distant parts of the Union. The period has passed when Congress hesitated to annex the territories of the Far West because of their great distance and the enormous time which would be required for a trip from those regions to Congress at

[1] It is divided into nine sections: A. Mathematics and Astronomy; B. Physics; C. Chemistry; D. Mechanics and Engineering Sciences; E. Geology and Geography; F. Zoölogy; G. Botany; H. Anthropology; I. Sociology and Economic Science.

Washington. But despite the great facility of travel in America — the continent may be crossed in four days — the formidable distances to be covered constitute a great obstacle in the way of scientific coördination, and now that the region of the Pacific is developing at a fast rate, it has a tendency to form its own associations and societies.

CHAPTER XIX

GENERAL CONCLUSIONS

LESSONS TO BE DRAWN FOR FRANCE. NECESSITY OF A
RENEWAL OF THE AGENCIES AND STRUCTURES
OF OUR INTELLECTUAL LIFE

> *" The world has been remade
> during the last half-century."*

Excess of "Stateism" in our university life. Rector or President? Liberty
as the condition of public support. Organization of student life. The societies of the friends of the universities; how to vitalize them. Excess of individualism among the students, the professors, and the instruction. Pure
and applied science. *L'École Polytechnique*, its organization and present
condition. Instruction and the organization of research. Institutes devoted
exclusively to research. The French universities must be greatly developed
after the war. They should be more varied and be mutually complementary
and not copy nor compete with one another.

We must renew our entire national structure, and not least, our scientific
organization the form of which no longer responds to the needs of today and
above all of tomorrow.

THE author desires to defend himself against the
possible criticism that he wrongly ascribes to his
readers a complete ignorance of university and scientific
life in the United States and that he is under the illusion
of having discovered America in this respect. He is perfectly aware that these matters are familiar to a certain
number of Frenchmen and that they have been treated
in more than one book. But judging from what he himself knew or was ignorant of before visiting the United
States, he thought it would not be unprofitable, and in
any case would clarify matters, to give a general account of the topic where the relation between the various parts would be made apparent.

The moment has now arrived for drawing conclusions from this account. So far as America itself is concerned, these conclusions have been indicated in the course of the various chapters, especially in the chapter which brings the first part to an end, and I content myself with calling attention once more to two general features which emerge from the facts already considered; first, the rapidity and amplitude of the recent growth of American scientific life; it is only now that the fruits of this movement will really ripen; and secondly, the great enlargement in the notion of the university which, advancing beyond our consecrated number of five faculties, covers at present all the branches of modern society where a profound intellectual culture is called for, and spreads the methods of positive science and the idea of its power everywhere and in generous measure.

I should like, on the contrary, to call attention to some lessons which I think can be drawn for the benefit of France from the facts already ascertained. A revision of all the elements of our national life is absolutely indispensable at the present moment, to the end of obtaining a better yield from our national organism after the war. A comparative study would supply a most solid basis for this purpose. Not that we should, purely and simply, introduce into our country institutions from the outside, whether American, or English, or German. Even had these latter been perfect, such a plan would have been none the less impossible. For their excellence lies above all in their relations with surrounding conditions, with traditions and customs of the countries whence they spring. But it might be useful to point out certain contrasts and to analyze them. Of course I can only just touch on certain points by way of illustration; to develop them would require a whole book.

The first and most significant of contrasts in the field of university life is that between our French *étatisme* and the vigor of private American initiative, a heritage of the English tradition. So far as higher education in France is concerned, hardly anything solid has been done — at least up to a very recent period — outside the State. Private higher instruction, whenever authorized by the law, has been dominated by political considerations that have vitiated it without giving it any real vitality. However, quite recently, private initiative has begun to produce some interesting institutions, such for example as the "*École Libre des Sciences Politiques*." In the field of scientific research, the *Institut Pasteur* above all, is a witness to what private initiative is capable of producing in France when accompanied by the benevolent support of the public.

But freedom will not be less fertile if granted more generously to the state institutions themselves. Our universities have been too closely subservient to and shackled by the meddling tyranny of central power in the greater part of their activities, even the most insignificant. We are not proposing to emancipate them completely. In the mechanism of our habits, the State is the only power sufficient to keep them alive. But the American example is suggestive in so far as it may contribute toward introducing a much stronger dose of initiative and autonomy into French university life.

The following seems to me a very significant fact: the way in which a French university is controlled or rather placed under close tutelage. An American university has a president at its head who is armed with considerable and perhaps too autocratic powers but who may and ought to use all his energy for the securing of the interests of the university, without having to take care of other interests as well. He is his own man. With us, the Rector who acts in the name of the university and presides over its Council does not issue from it at all. He is a simple functionary who receives his powers from the Ministry of Public Instruction, and who must administer the university on behalf of the State before he would ever dare not only to take any initiative on behalf of the university but even to defend its interests, in so far as these might differ from those of the State. Neither can he devote all his activity to the university for he carries besides the very heavy burden of secondary education, and to a certain extent, of primary education in his academy. That there should be a rector representing the State in an academy is very natural. But every university should have its own head who speaks its language and works for the realization of its projects, discussing matters with the rector on a footing of equality, while the rector himself should be a public minister. In Prussia, a country which is not generally regarded as the land of liberty, the State is represented in the university by a curator only; the rector is a direct and sovereign representative of the university itself. The same is true of other countries. Our rectors represent the survival of the entire Napoleonic régime and the complete subordination of higher education not even to the central authority but really to the central administration; the university is placed under the power of the bureaucracy.

The rector, unlike the president of an American university, cannot concentrate his mind upon the advancement of the interests of the university; he can only see to it that its aspirations are in accordance with the views which prevail in the offices of the *Rue de Grenelle*. It is only by the rarest exception that actually the University of Paris has as a rector the man who has been the principal reformer of higher education in France and who incarnates the university itself for us. M. Liard has, by his personal authority (even in opposition to the central administration), by his ability in affairs and by his devotion, served the University of Paris as a president after the American fashion. However that may be, we are directing our criticism not to any particular men but to a wholly illogical system. The only correct solution is that each university should have at its head a man who issues from it and entirely belongs to it. And his independence and facility of action will measure the degree of liberty and autonomy which the state will have granted to the universities.

Such a measure of liberty is an indispensable condition for the establishment of a real contact and confidence between the public and the universities and for the winning of the effective support of the former for the latter. If the private universities in America have been able to live and grow on a scale that has been indicated, thanks to the inexhaustible generosity of individuals, one of the principal reasons is that these latter are associated in the governments of the universities and control them to a certain extent. The state universities themselves, despite the forces which have brought them about, are beginning to provide a scope

for private action in their direction [1] and will enlarge
it no doubt in the future, because they will not allow
themselves to be deprived of the power represented by
the mass of their alumni.

The alumni! — there is a genuine point of contact
with the public, a point which we have done nothing to
create. The universities, clothed in the passive in-
sensibility of the state with respect to individuals, are
quite indifferent to those whom they have instructed.
They have made no effort to keep any trace of them,
much less to bring them back. Once the parchment
has been delivered by the universities, or rather by the
state through them into the hands of the candidates,
these latter become strangers, just as they were before
crossing the threshold of the university.

And similarly throughout the stay of the student in
the university, where could the least effort be discovered
for organizing his life, for establishing even then a link
between him and the university? Through a complete
misunderstanding of the psychological aspect of the
case, the authorities have suppressed all the ceremonies
and reunions which might have awakened in the student the idea of the academic community. Even then,
the student is ignored by the university; such culture
has with force and justice been called inhuman culture
by Mr. Barrett Wendell, whose judgment cannot be
suspected of malevolence.

The example of the American and the English universities should lead us to innovations in this respect,
not to any servile imitation, but to an adaptation of
new instruments to our habits so that we might concern
ourselves with the material condition of our students

[1] Consider for example the biological station of San Diego (Chap. XIV).

in order to improve it and to cease leaving them completely to themselves. The foreigners who will attend our institutions after the war, may, by importing their customs too, serve as a ferment and as guides to our students. It is desirable that we encourage them. And we should moreover help them to find in France an equivalent of what they have had at home, and to change the needs of our own students. Private initiative must be stimulated and encouraged to this end, for it is the only effective agent. But it must be able to rely upon the sympathy and support of the university itself.

When the universities were officially reconstructed twenty years ago, the need was felt of bringing them closer to the public. But the attempts made have suffered on the one hand from the inertia of the public, and on the other from the difficulty we feel in stripping ourselves of our customs of governmentalism. To this the history of the Societies of the Friends of the Universities bears witness. The earliest was founded at Lyons, distinguished in France as a city in which private initiative is particularly welcomed, and the society was perhaps the most vigorous of all. These societies should have taken rapid steps forward; instead, almost everywhere the initial effort weakened instead of gaining strength, and today the greater part of the members of these societies are professors, who are by definition, friends of the university, but who, according to a very accurate remark of one of my colleagues, endow the associations with an autophagous character.

However, the idea out of which they were born was excellent. The public should have been associated with them to a greater extent, but the societies failed to avail themselves of the factors best tending toward vitality. As it appears, the former students of the universities were completely neglected in the enlistment of members for the societies, and yet were they not preëminently suited as friends who ought to be attracted and kept? Before everything, the attempt should have been made to give to these societies a real character instead of letting them exist as abstractions. There is yet time to reform matters and these associations should be embodied in buildings which might serve as hospitable homes for the alumni, stripped of the stern austerity of the faculties themselves and helping every one to traverse the first stages of his career. The associations ought to bear some resemblance to the Harvard Clubs and similar clubs. If established on this basis, would not private initiative generously undertake to install them and animate them at the start? For once the inner and concrete life starts, they will develop by themselves, through the advantages which they would offer to their members and the memories which they would evoke in them.

On the other hand, the associations were not properly connected with the life of the university, and here governmentalism is to blame, for it is never inclined to give up the least bit of its authority. As it is, they have no regular share in the work of the Councils.[1]

[1] This is a question which concerns not only the universities but the whole fabric of public instruction. However, the idea is beginning to dawn that the Superior Council of Instruction must include citizens representing the principal social groups and not only the staff of instruction or the representatives of the administration. We must add, in extenuation of the form of existing institutions that the development of liberty was hampered by the clerical problem and by the necessity of defending the secular State which was attacked more in the field of university life than elsewhere.

Finally, so far as their internal life is concerned, these societies were often too much absorbed by the professional members of the universities when these should have served as discreet advisers without assuming control of affairs.

In a general manner, we should consider without fear the introduction of a certain amount of outside control in the administration of the greater part of the machinery of the universities. It is planned to establish institutes of applied science in the faculties of science and one of the interesting suggestions is to introduce into the councils of such institutes representatives of the industries interested in their prosperity.

The excess of individualism, not less than governmentalism, is one of the weak points of our university and scientific life, as it is indeed of the entire French community. This spirit of individualism is manifest everywhere; in the life of the students in the first place, in that it is solitary and knows almost nothing of activities undertaken in common such as those already noticed, that fill the life of the American student. The few existing associations are quite young and have scarcely passed the period of infancy. They cannot be too warmly encouraged. But the best way to create this sociability, that is so desirable, would be to organize student clubs (either for men or for women) pleasant and comfortable, and also useful by means of the advantages which they would provide in their character as associations. The Americans are going to set us an example by organizing in Paris, as they are now trying to do, the *American University Union*. Our French youths cannot help being urged in the same direction when they see with their own eyes how their friends organize their activities. In any case, the spirit of solidarity is not less strong in France than elsewhere among young people, but it must be awakened by a common life. Comradeship in the *École Normale* and the *École Polytechnique* is as much alive and intense as, if not more than, in any foreign university; it results from a life in common such as is absolutely lacking in the life of an average student in Paris or in any of our provincial universities.

Individualism is not less extreme among the teaching body. In the large cities particularly, the professors are too isolated from one another. There is no centre where they may meet and become acquainted in a field of activity outside their professional occupations. There is nothing among us to remind one of the *Faculty Clubs*, and *Colonial Clubs*, which create such an atmosphere of cordial spirit in American university life.

Finally, individualism prevails to excess in professional life itself. The freedom which is justly granted to a professor in the conception and carrying on of instruction leads to an excess such that each one goes his own way, ignorant of that of his neighbor. Coördination in instruction is gradually diminishing. Each chair is independent. The Faculty of Sciences in Paris, Darboux used to say when he was its dean, is a feudal body. The various professors live in their laboratories, somewhat as the barons of the Middle Ages used to live in their châteaus, without concerning themselves with one another and without joining their efforts together sufficiently with the intent of achieving a common result.

Scientific research proper accommodates itself to such habits, because in principle it is essentially individualistic. Still, it suffers a great deal in that it requires more and more an expensive and multiple apparatus which would be secured much more easily and completely by an association of efforts that would avoid any useless duplication of function. America has sometimes the reputation in Europe of being a country of prodigality and waste. It is sufficient to have seen the American university libraries — the general library of the university and the laboratory libraries — in order to perceive that there is infinitely more order and economy, and in consequence, an infinitely greater amount of utilizable resources in America than here. Without diminishing in any respect the liberty of each professor in his research work, it is imperative that the life of the various laboratories be better coördinated.

But this is truer still of the teaching, and above all of that teaching which is fundamental. Without that condition, a genuine instruction of students is impossible. Teaching must concern itself with the average person who needs to be directed methodically. Exceptional individuals are more or less able to dispense with guides, but after all they are a very small minority. And yet our system of superior instruction seems to have been made for these only. For such individuals are provided more liberally than elsewhere with higher and even unsurpassed courses, yet the very foundation on which this superior level should be established is far from being properly laid.

From the social point of view, power is obtained chiefly by the organization which strives to obtain plentiful returns of average value. Superior people escape all systems and triumph over difficulties and deficiencies; average people come to nothing if they are not systematically helped to reach a level where they may render important service to the community. The power of Germany for example, is due above all to a useful development and judicious utilization of individuals of average capacity.

We arrive at similar conclusions with respect to another field where the American example is equally striking; I have reference to the place of the applied sciences in the universities and chiefly of the engineering and agricultural sciences. Their importance is continually increasing, and not only do American universities assure the primary preparation of technicians in these fields but further they become centres of research through their laboratories and the special institutes which are attached to them; the Mellon Institute is particularly interesting in this respect.

In France, on the contrary, the universities have been conceived uniquely as instruments of pure science, and it is only very recently that they have turned more or less timidly toward the applied sciences, and moreover with quite insufficient resources. When the Revolution reorganized education, it isolated applied science within special schools access to which soon became very difficult owing to the introduction of competitive examinations. One of the consequences of this has been a profound anaemia in the faculties of science, from which almost the entire youth turned away; in fact, all the practical careers to which the study of science might lead were recruited outside the faculties. At least the preliminary theoretical instruction which is necessary for these professions should have been left to the Faculties of Science, even if we grant that the purely technical preparation might have been left in the hands of special schools. And yet, whenever instruction of this sort was required, it was organized specially outside the Faculties of Science. And now we face the paradox that the *École Polytechnique* in contradiction with its name and the purpose of its foundation has become a sort of a faculty of pure sciences, in which application properly speaking plays a very limited part.

The *École Polytechnique* is an institution without its like in other countries, one that has for a long time almost monopolized French science in a most brilliant fashion and which even today makes grave encroachments upon the scientific recruitment of our universities. The competitive examination which guards its gates exercises a powerful attraction upon youth and it is one of the essential factors in the great growth of our mathematical instruction. It results in a very severe selection, bringing to the school a student body of excellent quality. In fact, the strength of the school lies more in the quality of its students than in the curriculum.

The latter, as perpetuated by tradition is much to be criticized from the points of view both of pure science and of its applications. The courses are exclusively theoretical and chiefly mnemonic; one is even tempted to call them psittacizing when one recalls the system of repeated questions on which the students are solely judged. Success comes to the one who repeats most faithfully on the blackboard the literal content of courses hastily digested; it depends on the speed of the assimilation and on the physical resistance to a régime

which leaves almost no part to reflection or to original-
ity of mind and involves nothing of contact with reality
or experience. The system, if well applied, would suit
mathematics at most; but it is absurd to apply it to
such sciences as chemistry, and to cast them in the same
mould as analysis or rational mechanics, without
bringing laboratory work to bear.

How could an intellectual culture of this sort prepare
men, who are even well-endowed such as those generally
furnished to the *École Polytechnique* by the competitive
examination, to analyze reality? Their minds have
formed the habit of neglecting the observation of things
and of reducing the most complex of questions to a
small set of mathematical syllogisms. Of course, the
education of an engineer should embrace a considerable
portion of mathematics as an indispensable instrument,
but above all it must orient itself toward experimental
reality.[1]

Moreover, though a solid scientific instruction be useful as an introduction to very different applications, we
have however the paradox that it is a grading obtained
by a total adding of examination points which decides
the career that the graduating student will choose from
among the varied and heterogeneous group introduced
to him by the school, whereas the special aptitude of
the student does not influence the selection. It would
be quite reasonable, for example, that engineers specializing in explosives should be recruited preferably on

---

[1] From this point of view, mathematical instruction as given in our advanced schools of engineering and principally in the *École Polytechnique* is
on the whole too advanced and too theoretical. There is a confusion between that which is necessary to the masses and that which interests but
a small intellectual aristocracy.

the basis of their ability in chemistry.[1] This extreme presumed generality of interests is in conformity more with conditions as they existed when the *École* was founded in 1795 than with those today; and it is opposed to the spirit of specialization dominant in increasing measure in the foreign schools.

Finally, we should not forget that the annual selection of two hundred students for the *École* by competitive examination, though it may bring good material together, would immobilize for a period of two and often three years — the best years of youth in every respect — more than a thousand young men in tasks consisting of artificial exercises. So that however well the school might employ those who enroll in it, it has this to be said against it, namely, that to achieve this result, it risks the sterilization of two or three times as many individuals of whom many were about as able as those that have been received. How much better is the system of free admission into the universities, where abilities are manifested and classified, where tastes and aptitudes are formed and directed naturally to the appropriate special studies and where there is no pretension of stamping the twenty year old student in a final way and for all his life with the seal of this or that career.

The old ideas which separated the pure from the applied sciences and regarded only the former as worthy objects of study has the disadvantage not only of keep-

---

[1] Not many years ago, a considerable number of students used to drop out as the school finished and to utilize the knowledge they had acquired in following studies of their choice. This fact constitutes a serious argument in favor of the transformation of the *École* into an institution which would open its doors much more broadly, and which there would be no reason to separate, as is being done today, from the universities.

ing the mass of youth away from the universities, but of injuring science as such, in both its aspects, as pure and as applied. The first must come in contact with the second in order that it may make progress and if kept at too great a distance from the latter, it runs the danger of becoming a mandarinate. Moreover, pure science can suit a very limited number of minds, and the only practical method of discovering these minds is a system of free selection from among many individuals. In a *milieu* which is necessarily limited like that of a school of pure science, the conditions for such a selection do not exist. Great men are not made by education; the problem is only how to discover them in the large mass, how not to choke them, and how to secure for them a free development through the surest methods.

The most reasonable conception of a university, so far as science is concerned, is to secure a broad basis for the university by means of relatively elementary courses leading to varied and practical careers, and attracting the student public in a way which would permit a selection; then, at a higher level, the university should place advanced courses at the disposal of those who have been selected and chiefly an organization rendering research possible under favorable conditions. Above a certain level, instruction through courses *ex cathedra* is more or less futile. It is only the working in direct contact with reality that is fruitful. In this respect we are apt to consider advanced instruction in too absolute a fashion, as a process involving necessarily a professor. When the work of a scientist has received public recognition, if the public authorities decide to make a contribution in order to aid the scientist to go on with research and to stimulate others to the work of scientific investigation, they establish a new chair at the Sorbonne with its paraphernalia of oral courses and, invariably, examinations and diplomas. But almost the last thing they think of organizing is a laboratory, and never sufficiently, although a laboratory would have been the most necessary and urgent of things under the circumstances. We have a patent example in Pierre Curie, for whom, after his discovery of radium, a chair was established in the Sorbonne. But he died — prematurely it is true — without having the laboratory that was to him of all things the most indispensable.

In propounding this view before a Committee of the Senate some years ago, I had the opportunity to see that certain among those Members of Parliament that are interested in the questions of higher education are still far from understanding the distinction that I have just made. And yet, what I am suggesting is not something new; in all the large countries, the organization of scientific progress takes the form of the establishment of institutes devoted exclusively to research. France had shown the way a long time ago. The *Collège de France* and the Museum answer these conditions; however, oral instruction in them has been given too rigid a position, at least so far as the experimental sciences are concerned, and the laboratories have often been left in a lamentable condition. The *Institut Pasteur* is the model for all the great institutions of research that must be established when a new branch of science is being developed in the hands of a man of superior capacity. And there is no reason why they should be completely separated from the universities.

While we lingered among the old pedagogical formulae, in respect of the organization of scientific research, Germany was busy establishing on an ample scale in the years before the war the Institutes of the *Kaiser Wilhelm Gesellschaft*. And America witnessed the birth of the Carnegie and Rockefeller Institutes, and within the universities, the establishment of research laboratories such as the Wolcott Gibbs Laboratory of Harvard and a certain number of others which are proof that the value of the principle in question had been fully recognized.

Napoleon I had decided that the Faculty of Sciences of Paris, which in his eyes was chiefly a permanent committee of examinations, should have eight professors. Moreover, these professors were not to be in its own right; two were from the *École Polytechnique*, two from the *Collège de France*, two from the Museum and two from the Colleges. The Faculty of Letters had an analogous composition. For a long time, the number of chairs remained stationary, as is testified by the announcements of courses dating from the middle of the nineteenth century which have been made public in various expositions. But today things have improved; the chairs and the courses have been multiplied to a degree that has sometimes seemed scandalous to people occupying considerable administrative posts, as I have been able to observe. And yet our universities, especially those in the provinces, still remain of quite modest proportions when compared with foreign universities where it is not rare to find two or three hundred professors and as many, if not more, assistants who complete the teaching and make it accessible to the students in every branch of thought.

Our contemporaries have known the old Sorbonne; my generation, which is yet relatively young, has conducted its studies in the laboratories installed in the old dilapidated buildings which bordered the *Rue St. Jacques*. The new Sorbonne which is scarcely more than twenty years of age is a palace and looks like a whole world in comparison with the old buildings. But when we compare her with the size and the installations of the American universities, considered in this book, or with the large German universities, she seems rather small and — above all — choked and incapable of extension. What shall we say then of our provincial universities of which some at least should be able to bear comparison with the best foreign universities, and for the sake of the country's spiritual health, should equal that of Paris and counterbalance its influence! The Republic had done enormous good to higher education; in fact, it has practically created it. But no one should be deceived into thinking that it had a very great vision. The least trip outside France, the pilgrimage to Strasbourg that we all hope soon to undertake would suffice to undeceive us.

The present war has once more brought these problems to the foreground. The practical value of science has been confirmed more forcefully than ever as a source of power and wealth. Germany has drawn her aggressive audacity and above all her force of resistance less perhaps from the sickly exaltation of militarism than from the confidence in the resources which her scientific development assured to her. Where would she have been today if only her chemists had not realized the synthesis of nitrates in industry, indispensable as it is in the manufacture of explosives? And the fact that this was possible is due primarily to the prosperity of her universities.

These considerations are not relevant to us only. Certainly, England has taken them into account and is preparing, in its turn, to make up for lost time. Its universities, its laboratories, its technical schools based on modern ideas and above everything on the fertility of the experimental method, will make a considerable advance. Italy, now actively renascent, is no longer blind to these signs. If then our universities, instead of being revived, equipped with the necessary tools, and supported financially as they deserve, be left to remain on the morrow of the peace as they are today, before long we shall lag far behind those nations which aspire not to dominate the world but to live an independent life without being the satellites of those countries that will produce and inevitably regulate the condition of the others.

We should then think of developing our universities to quite vast proportions. What is aimed at is neither a luxury nor a chimera but a vital necessity. Yet, given the needs and the resources of the country, it is out of the question to attempt to establish fifteen gigantic and complete institutions. The effort must be concentrated on a smaller number. When the question of restoring the French universities was under consideration thirty years ago, the plan was to organize eight or nine large institutions only, in which it would have been possible to concentrate the existing resources. But owing to local interests which find such a strong support in the parliamentary régime, the plan failed, with the result that all the old groupings of faculties were transformed into universities and thus the effort was scattered. Doubtless, none of these universities is useless. They serve as centres of culture in the cities of the provinces which indeed need to be stimulated and stirred from the torpor into which excessive centralization in France has plunged them. But since the number of universities is rather too large, and the distances separating them often small, they should aim at supplementing one another, if they are to live in a genuine fashion instead of competing against and imitating one another, to the loss of all. If for example, Grenoble owing to its geographical position lends itself particularly to the development of such branches of study as electro-technics, it would be absurd if all the fifteen universities, following the example of Grenoble, tried to have institutes of electro-technics when four or five would be sufficient. Clermont-Ferrand can serve very well as a centre for the geological study of volcanic phenomena in connection with related facts such as the properties of mineral waters. Puy-de-Dôme is the seat of a meteorological observatory which has been set working in an interesting direction by its founder, but chiefly by my regretted friend, Bernard Brunhes. We have there the conditions for a great development of the study of meteorology which it would be futile to attempt to emulate elsewhere. Every region of France ought to stimulate the growth of some particular branches of science, for the purposes of which a university would serve as a metropolis attracting masters and students from afar. The universities have a definite rôle to play in the necessary awakening of the spirit of regionalism but on condition that they fit into the region and do not ignore the neighboring regions. The six hundred American colleges and universities are not and will never be institutions of equivalent significance; the universities, which today are in the foreground, will even tend to diminish in number and to become more diversified. Similarly, the destiny of the various French universities, if the circumstances are favorable, would seem to be to secure in every one the possibility of good fundamental studies, for which great equipment is not required. My own experience with my students who have come from different centres has taught me that often natural history is usually studied better in a modest faculty, like that of Besançon or Grenoble, than in Paris. The best condition is the presence of some good teachers, surrounded by a few students, and animated by the sacred fire. But for the purposes of specialization, a perfect and rich apparatus is necessary; there as elsewhere, division of labor and coördination are imperative, and the spirit of imitation and sterile competition should be avoided.

For anyone who returns from America to France, an impression which is not very pleasant but is very persistent accompanies the perception — reënforced by comparative observation — of the fineness and the profound virtues of our ancient race; it is the impression that our national fabric, intellectual as well as economic, is scanty and oldish. Our institutions were brilliant and fruitful a century ago; they were then ahead of their time. But we have remained content with our past glory without adapting ourselves sufficiently to the new conditions. *The world has been remade during the last half-century* and yet, in many respects, we are still cast in a mould which was fitting enough to the period of Louis-Philippe. It is better to admit this frankly and to analyze its causes than to delight in the illusions of a *bourrage de cranes* — to use an expression of the day — duping only ourselves.[1]

This is true generally and not by reference to any particular aspect of social life; and it originates from very profound causes. Without taking sides in a political dispute, I should say that it is the result of bourgeois mentality. The French *bourgeoisie* which has been in control for a century has encrusted itself in some way with genuine virtues that unfortunately are of secondary value and kill vitality. A biologist would be tempted to compare this condition to an encystment or to a similar form of lethargic life or even to the life of the organisms trained to live in aquaria by means of a reduced diet and diminished nutritive reactions, which however lack the liveliness and the fertility of their free fellow-creatures.

Its ideal has been to preserve the wealth already acquired, working to this end for the greatest possible security, turning away from adventure and the life of enterprise with its possibilities of loss but also with its

[1] After writing the above, I met a former student of mine who is a foreigner, ripened by hard experience and long personal work, and who after having obtained all his scientific education in France (to which, by the way, he remains extremely attached) settled in England, two years ago. Quite spontaneously, he made many remarks to me, which were the fruit of his observations and which I had already been led to record here, and he asked me with sincere anxiety if France after the war would be sufficiently aware of the need to modernize all her life — a need which is so apparent to anyone who has lived in foreign countries recently. The conversation was for me a confirmation of the observations which I have brought forward here and a proof that they are in no way exaggerated.

chances of success. It even scorns the occasions which are offered, as the recent history of our colonial rule proves. Instead of putting its savings into investment for the purpose of augmenting the wealth and the strength of France, it permits its banks to serve as sleeping partners in foreign enterprises and to arm other peoples who are thus enabled to procure weapons with which to attack us.

Likewise, it has voluntarily relinquished the task of propagating the race in order to save itself from the trouble of producing new wealth, thus committing collective suicide. The people imitate and by indulging in the taste for comfort, expose the country to the most terrible danger of the hour.

As a consequence, the *bourgeoisie* has been too indifferent to whatever contributes to the renewing of the environment and everybody has been trying above all to maintain the state of things already realized without noticing that the latter, like a position turned by the enemy, falls to pieces by the mere changing of the external situation. The French *bourgeoisie* as a whole has been unaware of this fact, for it scarcely traveled at all. It thus saved the expense of the trip. I have more than once had this feeling expressed to me in conversation with people in the cities of the North, rich or poor. "What have you gained," they used to ask me, "by making such a long and expensive trip just to be present at a congress?"

The same state of mind is responsible for the fact that the French public remains deaf to all appeals inviting it to take the initiative in tasks of public interest; it lacks the private initiative which in the Anglo-Saxon countries is ever awake, ever sure of the generous aid of the richer classes and is the source of the fundamental idealism of these peoples, despite the utilitarian picture which we paint of them. The French *bourgeois* prefers to let the State do whatever is necessary, while at the same time insisting on paying the smallest taxes possible and is little concerned about using the resources placed by the State at his disposal.

He is far from lacking culture; but, during the whole of the nineteenth century, his culture has been too exclusively literary, abstract and formal. It has given rise to a genuine finesse, an incontestable elegance of mind and has safeguarded the qualities of high-minded sentiments, bravery in danger and the broad feeling of human solidarity which reappear with all their force at the moment of a great crisis; the present war has furnished a magnificent proof of this. Yet a culture of this sort is not adequate to the conditions of modern life; notice also the fact that literature, in seeking to renew itself indefinitely, after having more or less exhausted the analysis of human nature, now begins to turn to the field of the more and more exceptional, thus gradually slipping into pathology. The limits of art are becoming indefinite, and literature (even when dissociated from the interloping productions wrongly attributed to us and of which we were ignorant because these works were the product of foreign factories) has a rather unhealthy tone which shocks the stranger. We should not be surprised at this. The success we have obtained by this form of literature suggests the thought that it depicts our ordinary life, when really it is remote from the prosaic wisdom of the French masses.

But above all, the French public has been led away from interest in scientific culture by an excess of literature. It has been indifferent to science; it has had no faith in its power. In alleviation we should say that the Catholic Church — of which we must never overlook the educative influence — has been indefatigable in its efforts to cast suspicion on science; and even today, it is not averse to hearing its failure proclaimed. In the meantime, others at our side have been preaching to a whole nation as a fundamental axiom — and chiefly by the channel of the universities — the principle of the sovereign importance of Science as a factor of wealth and power. In fact, the importance and practical bearing of Science are far from being limited to the immediate consequences of the discoveries. Fundamentally, the scientific spirit controls the whole material aspect of social life to an increasing extent. As M. E. Picard — a mathematician devoted to very speculative researches, extremely remote from common reality — has very judiciously observed, the scientific spirit is in no way a particular entity by itself, but very simply a continuation of good sense. When applied to practical life, it is only the reasoned and absolute faith in the logical connection of facts and the rational prediction of effects from their causes. It is thus the antithesis of the ancient religious belief in miracles — and in the capacity of a supernatural intervention to modify things as we desire. Not less is it opposed to that attitude, derived in the main from the preceding one, according to which it is in no way necessary to concern oneself with remote previsions. One aims always to settle things when confronted by events; one counts on chance and finds one's way out; this is system D in the language of the trooper. This system may not be devoid of elegance; it may enable one to emerge from difficult situations, ingeniously and sometimes heroically but it is never useful for the purpose of building the structure of the future. We have an unfortunate national inclination toward it. We must react vigorously against it to the advantage of the scientific spirit which foresees and organizes and from which our adversaries of the moment draw their greatest power. The scientific spirit, thus understood, enters into the daily practices of the material life of a people in proportion as its methods of production, in all fields, are sane and fertile. There can be no assurance of social prosperity in modern life where this principle is not appreciated; and in the competition among the peoples, the decisive factor is perhaps the degree in which it has penetrated the mind of the public and has entered the domain of the unconscious. Scientific culture and above all the scientific spirit which comes from it and which remains identical at its various levels are thus factors of capital importance in the formation of the social outlook.

Political centralization — cultivated even today by the parties in power, as an instrument of control with almost as much relish as by Napoleon I — has been added today to the preceding causes in paralyzing the life of the provinces. Our average provincial cities — and even our large cities — make a painful impression on anyone coming from English or Swiss towns or from elsewhere. All this must be modernized.

The growth of our intellectual institutions must necessarily be bound up with the average intellectual development of our ruling classes. Control of the latter is an indispensable factor. People who, at a given moment, feel the need of progress and endeavor to bring it to pass, are helpless if not supported by public opinion. Under Napoleon III, Pasteur, Claude Bernard, Wurtz, Sainte-Claire Deville, and others sounded the alarm in the clearest way, but their appeals fell on deaf ears. Nowadays, in spite of the very important progress accomplished under the Third Republic, our academic institutions have still something antiquated about them which reflects the state of mind of the public, and for which the latter is to a great extent responsible.

Despite their new name, our universities have not yet stripped themselves of the spirit, the structure, and the chains of the Napoleonic faculties. The *Collège de France* has neither the laboratories nor the resources which it deserves. The Museum of Natural History, in spite of its souvenirs of Lamarck, Cuvier and Geoffroy Saint-Hilaire, periodically invoked, is not the Museum with which Paris ought to put in comparison with the British Museum, the American Museum, and other large foreign Museums. The wealth of the past is not sufficient to assure to it the rank which it ought to hold. Its library, which is so rich and precious because of its age, is not as well furnished as it should be, and the inadequacy of its means is not reassuring as to the safety of the riches which it does contain. The *École Polytechnique* despite the prestige which its uniform possesses in the eyes of the French middle class and the social force of the comradeship which it engenders, is an anachronism in many respects in modern higher education, as I have already pointed above. The contrary would have been surprising if one considers the fact that the school has, so to speak, not changed for a century, and it is an utter anomaly that even today the

Ministry of War determines the destinies and the régime of a school of engineers! The Institute too needs a rejuvenation in spite of the favor which the general public accords to it.

We must recast all these institutions each one of which has had its share of glory and preserves its intrinsic virtues, but altering and coördinating them anew so as to adapt them to present needs. At the same time, we must endow them all much more considerably than at present. Living is becoming more expensive for the scientific institutions as for individuals and at a much faster rate. The community, that is to say, the state must understand this. Public-spirited individuals of wealth and culture must help these institutions to be equal to their tasks. The richer classes of America offer a magnificent example in this respect, thus furnishing some excuse for the plutocratic régime which is sometimes justly condemned by democrats.

What is most important at the present is to realize that great effort will be called for after the war. Our race has sufficient resources to justify confidence in its future. In the course of these three years, France has demonstrated that she is capable of immediate and continued effort, whereas in the opinion of many strangers, judging from appearances, she was only the deposit of a past glory of which only the delicate but impotent charm had remained.

The battle of the Marne has been defined as a strategic restoration by the one who by winning it saved France and the liberty of the world. But it has also been the signal for a moral restoration of the whole world. Those who like me were in the United States in 1916 during those still doubtful but heroic moments of Verdun were able to estimate what France had regained in the opinion of the world and the credit she had won for herself in the future.

This credit will lapse after a short interval once the crisis is passed, and when each people has returned to its task, competition will inevitably commence once more with new bitterness. But France must maintain herself at the level where the events have placed her, since August 3, 1914; for this, she must pay the price of considerable effort, which will be the more arduous as the number of those participating in the struggle will have singularly diminished, and diminished by the loss of the best units.

It will therefore be urgently necessary for France to secure maximum results from a given effort, in every field. France must apply in a judicious manner the principles developed by F. W. Taylor in all the spheres of her national activity. To this end, she must achieve an economic and intellectual recovery, not less necessary than the strategical recovery of the Marne. She must modernize — resolutely and methodically — all her institutions, including the scientific, paying no consideration to the inertia of conservative gerontocracies.

I should advise those who want to understand matters better to visit the United States for a few months. There they will realize not merely the necessity of recasting our implements of work and action in modern moulds but they will also be encouraged by the admiration provoked by the effort of France and impressed by the idealism which goes into the practical spirit of the American people and which drives universities and scientific establishments at such a rapid rate in the path of development and progress.

It is to this that the little book in hand attempts to bear witness; it shall end by asserting once more the profound importance of developing the mutual acquaintance between the men of science and the academic institutions of the two countries. Such an acquaintance generates not only sympathy but also power.

APPENDICES

I. STATISTICS OF THE STUDENTS IN THE PRINCIPAL UNIVERSITIES IN 1914–1915
(Quoted from *Science*, December 25, 1914)

	Chicago	Columbia Univ. including Barnard College	Cornell	Harvard and Radcliffe College	Johns Hopkins	Pennsylvania	Princeton	Stanford	Yale	California	Illinois	Michigan	Minnesota	Wisconsin
College —Men	911	1014	926	2479	294	438	1327	663	1437	1238	505	1802	816	871
—Women	746	689	279	603		18		427		1853	426	780	905	874
Graduate School of Arts and Sci.	598	1689	321	512	230	489	175	130	371	478	340	258	169	321
Theology	152			59					112					
Law	213	440	235	716		356		177	142	134	112	499	171	168
Medicine	200	358	151	321	374	290		73	50	128	287	378	213	96
Dental Schools				204		663				112	84	318	253	
Pharmacy		495								95	199	110	97	32
Pedagogy	1262	1817				89				*			94	46
Fine Arts			157	*					38	*	*	*		
Architecture		110				*				16	68	145	*	84
Music		3				22			82	*	*	*	*	101
Journalism		136										*		469
Commerce	170			147						287	376	*		796
Engineering and Applied Sci.		461	1363	120	*	1615	139	418	1056	763	1406	347	590	1091
Agriculture			1535			906				540	959		598	
Forestry Schools									37			*	37	
Veterinary Schools			116			122								
Miscellaneous	881				160	743					379		15	41
Total †	3887	6752	5078	5161	1058	5736	1641	1888	3289	5614	5137	5522	3940	4874
Summer Schools	3983	5590	1436	1250	356	983		14		3179	838	1594	867	2602

* Included under other headings.   † Double registrations have been deducted.

II. INCOME OF THE PRINCIPAL UNIVERSITIES (1913-14)

(Rep. Comm. of Education)

NUMBER OF PROFESSORS AND ALUMNI

Institutions	Income from Students	Income from Endowments	Total Income (gifts, etc. included)	Budget of Expenses 1912–13 *	Number of Professors †	Number of Alumni in 1909 †
Chicago	900,000	1,082,000	3,332,000	1,550,000	291	4,915
Columbia	1,660,000	1,287,000	7,892,000	2,176,000	559	18,000
Cornell	580,000	610,000	6,790,000	1,752,000	507	9,350
Harvard	900,000	1,345,000	4,288,000	2,347,000	573	19,000
Johns Hopkins	121,000	244,000	738,000	372,000	172	2,000
Pennsylvania	693,000	200,000	1,686,000	......	375	15,000
Princeton	373,000	266,000	1,433,000	710,000	163	6,175
Leland Stanford Jr.	100,000	925,000	1,045,000	......	136	2,800
Yale	742,000	810,000	2,600,000	1,324,000	365	15,500
California	145,000	144,000	2,500,000	1,840,000	350	7,950
Illinois	240,000	......	2,825,000	2,000,000	414	6,600
Michigan	428,000	19,000	2,202,000	1,423,000	285	20,200
Minnesota	360,000	......	3,034,000	......	303	7,200
Wisconsin	573,000	8,000	3,102,000	2,100,000	297	6,750
Mass. Institute of Technology	385,000	120,000	694,413			
Bryn Mawr College	260,000	75,000	355,000			
Smith College	501,000	123,000	1,005,000			
Vassar College	561,000	75,000	1,107,000			
Wellesley College	625,000	70,000	1,466,000			

* According to Minerva.   † Slosson, *loc. cit.*

PRINTED AT
THE HARVARD UNIVERSITY PRESS
CAMBRIDGE, MASS., U. S. A.

HISTORY, PHILOSOPHY AND
SOCIOLOGY OF SCIENCE

Classics, Staples and Precursors

An Arno Press Collection

Haldane, J. B. S. **Science and Everyday Life.** 1940

Hall, Daniel, et al. **The Frustration of Science.** 1935

Halley, Edmond. **Correspondence and Papers of Edmond Halley.**
 1932

Jones, Bence. **The Royal Institution.** 1871

Kaplan, Norman. **Science and Society.** 1965

Levy, H. **The Universe of Science.** 1933

Marchant, James. **Alfred Russel Wallace.** 1916

McKie, Douglas and Niels H. de V. Heathcote. **The Discovery
 of Specific and Latent Heats.** 1935

Montagu, M. F. Ashley. **Studies and Essays in the History of
 Science and Learning.** [1944]

Morgan, John. **A Discourse Upon the Institution of Medical
 Schools in America.** 1765

Mottelay, Paul Fleury. **Bibliographical History of Electricity and
 Magnetism Chronologically Arranged.** 1922

Muir, M. M. Pattison. **A History of Chemical Theories
 and Laws.** 1907

National Council of American-Soviet Friendship. **Science in
 Soviet Russia: Papers Presented at Congress of
 American-Soviet Friendship.** 1944

Needham, Joseph. **A History of Embryology.** 1959

Needham, Joseph and Walter Pagel. **Background to Modern
 Science.** 1940

Osborn, Henry Fairfield. **From the Greeks to Darwin.** 1929

Partington, J[ames] R[iddick]. **Origins and Development
 of Applied Chemistry.** 1935

Polanyi, M[ichael]. **The Contempt of Freedom.** 1940

Priestley, Joseph. **Disquisitions Relating to Matter and Spirit.**
 1777

Ray, John. **The Correspondence of John Ray.** 1848

Richet, Charles. **The Natural History of a Savant.** 1927

Schuster, Arthur. **The Progress of Physics During 33 Years
 (1875-1908).** 1911

Science, Internationalism and War. 1975

Selye, Hans. **From Dream to Discovery: On Being a Scientist.**
 1964

Singer, Charles. **Studies in the History and Method of Science.**
 1917/1921. 2 vols. in one

Smith, Edward. **The Life of Sir Joseph Banks.** 1911

Snow, A. J. **Matter and Gravity in Newton's Physical Philosophy.** 1926

Somerville, Mary. **On the Connexion of the Physical Sciences.** 1846

Thomson, J. J. **Recollections and Reflections.** 1936

Thomson, Thomas. **The History of Chemistry.** 1830/31

Underwood, E. Ashworth. **Science, Medicine and History.** 2 vols. 1953

Visher, Stephen Sargent. **Scientists Starred 1903-1943 in American Men of Science.** 1947

Von Humboldt, Alexander. **Views of Nature: Or Contemplations on the Sublime Phenomena of Creation.** 1850

Von Meyer, Ernst. **A History of Chemistry from Earliest Times to the Present Day.** 1891

Walker, Helen M. **Studies in the History of Statistical Method.** 1929

Watson, David Lindsay. **Scientists Are Human.** 1938

Weld, Charles Richard. **A History of the Royal Society.** 1848. 2 vols. in one

Wilson, George. **The Life of the Honorable Henry Cavendish.** 1851